American Institute of Chemical Engineers

Topical Conference On Distillation 2005

Held at the 2005 AIChE Spring National Meeting

April 10-14. 2005
Orlando, Florida, USA

Printed from e-media with permission by:

Curran Associates, Inc.
57 Morehouse Lane
Red Hook, NY 12571
www.proceedings.com

ISBN: 978-1-60423-465-7

American Institute of Chemical Engineers

Topical Conference on Distillation 2005

Held at the 2005 AIChE Spring National Meeting

AIChE

April 10-14, 2005
Orlando, Florida, USA

American Institute of Chemical Engineers

Topical Conference on Distillation
2005

TABLE OF CONTENTS

Session 85 - New and Exciting Packing Applications

Chair: Henry Kister
Vice Chair: Mike D. Pritchett

Session 86 - Distillation Troubleshooting (The Dick Harris Memorial Session)

Chair: Henry Kister
Vice Chair: Mike D. Pritchett

81a Thermo Savvy for the Distillation Expert

Kenneth R Cox

This paper introduces new tools for the distillation expert who needs to further develop their understanding of thermodynamic property methods used in process simulators. This is not a paper directed at experts in thermodynamics, but rather toward the distillation expert who needs simple measures to determine when the phase equilibrium property methods require special attention in their distillation designs.

We often take it for granted that physical property methods in process simulation packages provide us with accurate representations of the real world. On the face of it this makes good sense considering that the physical property correlations are thoughtful fits of generally well accepted literature data. Most of the time these correlations will provide reasonable representations. Our focus is on the other cases where the correlations contribute to inaccuracies in distillation design. Our approach is to teach how who do not consider themselves experts in solution thermodynamics how to diagnose these situations and how to go about resolving the problem.

We present a set of quantitative measures and decision diagrams that will help the nonexperts in thermodynamics to determine when to pay special attention to the thermodynamic packages and chart a path to a accurate thermodynamic description. We give several example situations highlighting special challenges, such as pressure-swing distillation, near azeotropic separations, and separations involving volatile electrolytes. As a part of this discussion, we will review general guidelines for selection of a proper thermodynamic correlation and steps to be taken when existing correlations prove indadequate.

New Tools for Troubleshooting Distillation Simulation

Stanislaw K. Wasylkiewicz

Aspen Technology, Inc.
900, 125 - 9th Avenue SE, Calgary, Alberta T2G 0P6, Canada

Presented in the Distillation Symposium of the 2005 Spring AIChE Meeting,
Atlanta, GA, April 10-14, 2005, Paper No. 81b

Abstract

For rigorous, high fidelity azeotropic or reactive distillation simulation, robust thermodynamic models are extremely important. Although an outstanding progress in modeling of physical and chemical equilibrium has been made in the last century, there is no universal model that can be used for any mixture, at any conditions, without adjusting various empirical parameters. Furthermore, because we often neglect some, as we think, insignificant terms in thermodynamic models, adjustable empirical parameters have to account for not only what they were supposed to describe but also for all effects that have been neglected. This makes the models less predictable and their extrapolation to multicomponent mixtures questionable. Even if model predictions for all binary sub-mixtures are consistent with experimental data, we have to verify behavior of the whole mixture before thermodynamic packages can be used for rigorous simulations. Recently several new excellent tools have been developed for proper selection of thermodynamic models, their verification and troubleshooting. Phase diagrams, residue curve maps (RCM) and distillation region diagrams (DRD) are invaluable for this purpose because of their visualization capabilities. However, they are restricted to ternary and quaternary mixtures. There are no such restrictions for the azeotrope pressure sensitivity analysis (APSA), where temperature and composition bifurcation diagrams can be created for mixtures of any number of components. Since pressure changes from stage to stage in distillation columns, APSA has been proven to be an excellent new tool for quick model verification and troubleshooting of real, non-isobaric distillation columns. It reveals qualitative separation space changes difficult to find by RCM of DRD generation. In the paper, several case studies are presented that describe frequently encountered modeling problems during simulation of non-isobaric azeotropic distillation columns. Then the application of the new tools in troubleshooting of these simulations is described in details.

Keywords: azeotropic distillation, troubleshooting, simulation, separation systems

Introduction

A rigorous and reliable thermodynamic model is crucial for synthesis and design of systems for separation of azeotropic mixtures. Incorrect prediction of azeotropes in a mixture often leads to generation of completely false, infeasible separation sequences. Although an outstanding progress in modeling of phase equilibria has been made in the last few decades, there is no universal model that can be used for any mixture, at any conditions, without adjusting numerous empirical parameters. Moreover, because we frequently neglect some terms in the rigorous thermodynamic models, the adjustable empirical parameters have to accommodate not only what they were supposed to describe (e.g. interaction between molecules in liquid phase) but also all effects that has been neglected (Agraval et al., 2001). This makes the models less predictable and their extrapolation to mixtures containing more then two components questionable. Even if the experimental vapor-liquid equilibrium (VLE) data for all binary pairs are available and are used for regression of binary interaction parameters, we have to verify behavior of the whole mixture before any attempt can be made to synthesize separation schemes for these mixtures.

Because of their outstanding visualization capabilities, residue curve maps (RCM) are very useful for characterization of ternary and quaternary azeotropic mixtures. A shape of the separation space, including distillation boundaries, distillation regions and azeotropes (singular points of the maps) can be easily examined on RCMs (Doherty and Malone, 2001). This makes RCMs the excellent tool for selection and evaluation of thermodynamic models for ternary and quaternary mixtures. Villiers et al. (2002) show how the visual analysis of the structure of an RCM provides valuable insight and help to evaluate the underlying thermodynamics, and how it is useful to screen binary VLE parameters based on alternative databanks and regressions strategies. Because RCMs depict VLE behavior in a ternary composition space, they can be more relevant than the binary phase diagrams in assessing the validity of various parameter sets for distillation modeling. Villiers et al. (2002) showed a few examples of troubleshooting phase equilibria that include using data outside their intended range, employing VLE calculations instead of VLLE (vapor-liquid-liquid equilibrium), using parameter sets optimized to different regression objectives and choosing the appropriate model for the vapor phase. Wasylkiewicz and

Shethna (2002) presented other examples of utilizing RCMs and VLLE diagrams for selection of appropriate thermodynamic model and verification of interaction parameters. Wasylkiewicz (2004a, 200b) explained how these methods can be applied in troubleshooting distillation columns. Conveniently, RCMs can now be readily generated using commercial process simulation or synthesis software, such as Aspen Plus or Distil. Moreover, the singular points of the maps can be calculated for any number of components and used for testing of phase equilibrium models.

Azeotropes and their types determine distillation regions and distillation boundaries. Therefore, knowledge of temperatures and compositions of all the azeotropes in a mixture at a specified pressure is critical for both the design and operation of distillation separation systems. Strong non-linearities encountered in vapor-liquid-liquid equilibrium and the presence of multiple solutions, both real and spurious, complicate the task of computing temperatures and compositions of all the azeotropes in a heterogeneous mixture at particular pressure. One can attempt to find all of the solutions by using a nonlinear solver started at several starting points. However, even with extreme calculation effort, this approach does not guarantee that all of the azeotropes will be found. In the last decade, several new techniques have been proposed to solve this problem, e.g. the Levenberg-Marquardt algorithm (Chapman and Goodwin, 1993), the interval Newton method (Maier et al., 1998), a global optimization method (Harding et al., 1997) etc. The most reliable and robust method for the calculation of all the azeotropes in a homogeneous mixture has been proposed by Fidkowski et al. (1993) and later generalized to include heterogeneous liquids (Wasylkiewicz et al., 1999). This method together with an arc length continuation and rigorous stability analysis (Wasylkiewicz et al., 1996) gives an efficient and robust scheme for finding all of the homogeneous and heterogeneous azeotropes predicted by a thermodynamic model at a specified pressure.

In the real-world, the plant operation is not static. Changes in product quality and quantity often require operation at different pressures. Even when operating in a near steady-state condition, many columns exhibit significant pressure gradients between the distillate and the bottoms. Thus, determination of the pressure sensitivity of the azeotropes in a mixture can be critical for reliable design or troubleshooting of distillation separation systems. When upper and lower pressure limits are specified, the analysis can start from one pressure limit and than the azeotrope composition can be calculated for gradually increasing or decreasing pressures. This simple parameter continuation procedure can follow any azeotrope we already know. However, it can not find any new azeotrope, which appears as the pressure changes. Another solution to the problem could be the application of the rigorous azeotrope calculation procedures at each pressure interval. Such an approach would be extremely computationally intensive and time consuming. To overcome these difficulties we have developed a new azeotrope pressure sensitivity analysis (APSA) method (Distil, 2003). The method reveals how the compositions and temperatures of azeotropes change with pressure. It applies bifurcation theory together with an arc length continuation to do the calculations (Wasylkiewicz et al., 2003). This allows not only to follow individual azeotropes but also to find any new azeotrope that appears in a specified pressure interval. Therefore *all* bifurcation pressures, where azeotropes appear or disappear, can be found. The computation time required for the whole pressure sensitivity analysis is similar to the time necessary for the homotopy-continuation method to find all azeotropes at a single pressure.

The sensitivity analysis of azeotropes to changes in operating pressure provides new opportunities in the design of azeotropic distillation sequences. Increasing or decreasing operating pressures in individual columns changes composition of azeotropes and positions of distillation boundaries and even can cause their appearance or disappearance. This can have enormous effect on the topology of the residue curve map and the feasibility of distillation sequences. APSA shows opportunities for pressure swing distillation and heat integration between columns in the sequence. Pressure swing distillation can often be considered as an attractive alternative for breaking homogeneous azeotropes and sometimes can considerably simplify complex separation systems. The information about pressure sensitivity of azeotropes can be of critical importance in the design and troubleshooting of real distillation columns, especially when there is a substantial pressure drop in the column. In some cases, this can lead to a switch in topology of distillation regions inside the column and cause serious problems in convergence of simulators in steady state and dynamic modes. Examples of this quite common problems with simulation of azeotropic distillation columns are shown in the next paragraphs and the application of APSA for troubleshooting of these cases is described in details.

Troubleshooting Non-Isobaric Extractive Distillation

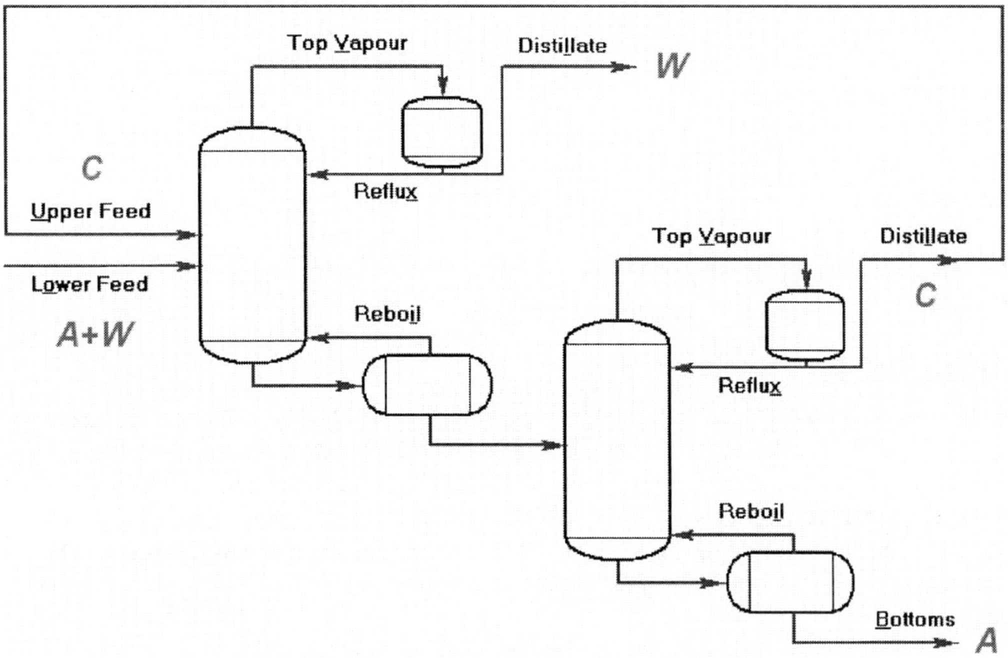

Figure 1. Breaking binary azeotrope AW in extractive distillation.

Let us consider a typical separation problem. A mixture containing mostly components A and W should be separated. There is a binary azeotrope between components A and W. Since the azeotrope is not sensitive to pressure change (see Table 1), an entrainer C was selected to separate the binary mixture in an extractive distillation sequence of two distillation columns (Figure 1). The first column was set up at constant pressure 150 kPa and converged quickly to desired products: high purity A and a binary mixture of W and C. Then a pressure drop in the column was taken into account. This time, the required high recovery of component A in the bottom of the column was never achieved even with an extreme reflux and a large number of stages.

Table 1. Temperatures and compositions of binary azeotrope WA at selected pressures.

Pressure, kPa	Temperature, °C	Mole Fraction of W	Mole Fraction of A
50	80.64	0.960958	0.039042
100	98.79	0.955925	0.044075
150	110.44	0.952532	0.047468

For the three key components (W, A and C) a distillation region diagrams (DRD) ware created for three selected pressures to examine the three-component space for azeotropes and distillation boundaries as shown in Figure 2. By examining the DRDs one can easily conclude that there is impossible to achieve complete recovery of component A in the bottom product of the first column at 50 kPa. This is because of a presence of a distillation boundary at this pressure. But in our simulation case, pressure at the top of the column does not drop as low. It always stays above 100 kPa. To troubleshoot this case further we use a new tool - azeotrope pressure sensitivity analysis (Distil, 2003).

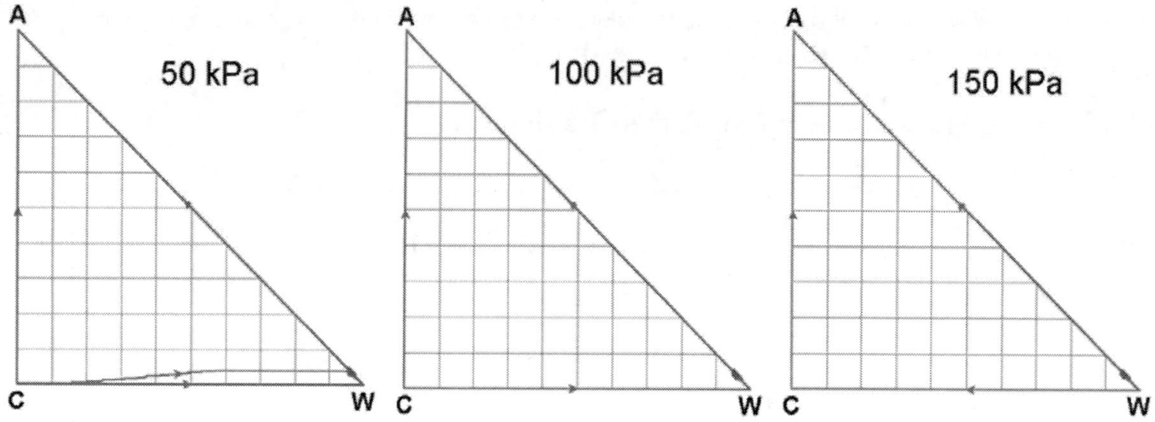

Figure 2. Distillation region diagrams for ternary mixture W-A-C at three pressures 50, 100 and 150 kpa. NRTL-Ideal model.

Azeotrope Pressure Sensitivity Analysis

The sensitivity of azeotropes to changes in pressure was discovered in XIX century (Roscoe and Ditmar, 1859). Since then, many theoretical and experimental studies were performed on this effect (see e.g. Roscoe 1860, 1862; Horsley, 1947; Zawisza, 1961, Malesinski, 1965, Zharov and Serafimov, 1975; Knapp, 1991). The magnitude of pressure effects depends on a mixture. Sometimes compositions of azeotropes change very little (see e.g. ethanol – water azeotrope in Wasylkiewicz et al., 2002). However, there are mixtures where compositions of some azeotropes change rapidly with pressure and even azeotropes appear or disappear as pressure varies.

Knapp (1991) introduced the concept of a *bifurcation pressure*, where an azeotrope appears or disappears, and showed necessary conditions for a homogeneous binary azeotrope to bifurcate. There is an azeotrope on one side of a bifurcation pressure and a tangent pinch on the other side. As the pressure increases or decreases away from the bifurcation pressure the severity of the tangent pinch decreases. Therefore a distillation column should not operate near a bifurcation pressure. In real distillation columns, pressure changes from stage to stage. This can lead to a switch in topology of distillation regions and cause serious problems in both real operation and in convergence of calculations for simulators in steady state or dynamic modes. In such cases, information about the bifurcation pressures could by extremely useful in troubleshooting both for simulators and real operations.

The azeotrope pressure sensitivity analysis method has been developed by Wasylkiewicz et al. (2000) and is based on bifurcation theory together with an arc length continuation. The method finds *all* bifurcation pressures in a specified pressure limit and allows not only to follow individual azeotropes, but also to find efficiently any azeotrope that appears in the pressure limit. For details see Wasylkiewicz et al. (2003). Here we apply this tool to create various bifurcation diagrams, which are extremely useful in troubleshooting of azeotropic distillation columns.

In pressure sensitivity analysis of azeotropes, we start a branch for each pure component and each azeotrope found at 50 kPa. Each branch is followed and continually checked for new bifurcations between 50 and 150 kPa or until the branch disappears. The corresponding homotopy continuation branches are shown in Figure 3, temperature branches in Figure 4 and composition branches in Figure 5. The binary azeotrope WA is not sensitive to pressure change (see also Table 1). On the other hand, the WC and WAC azeotropes are very pressure sensitive and exist only in a narrow pressure range (WAC between 89 and 127 kPa, WC only in the proximity of 127 kPa). There are two bifurcation points where the WAC branches collaps: one on branch WA (at 112 kPa), second on branch WC (at 127 kPa). There is also a turning point at branch WAC at 89 kPa.

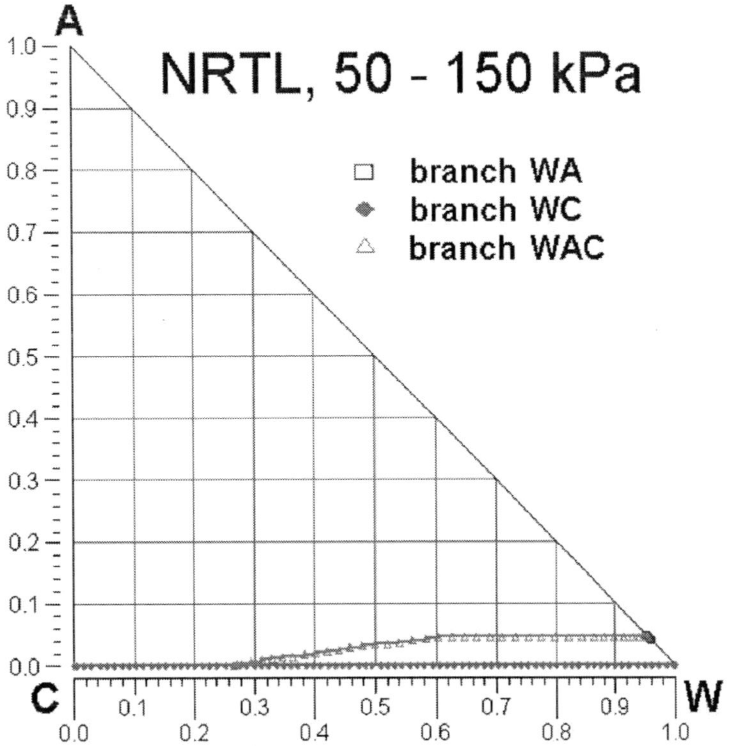

Figure 3. Homotopy continuation branches WA, WC and WAC for pressure analysis of azeotropes for ternary mixture W-A-C between 50 and 150 kPa. NRTL-Ideal model.

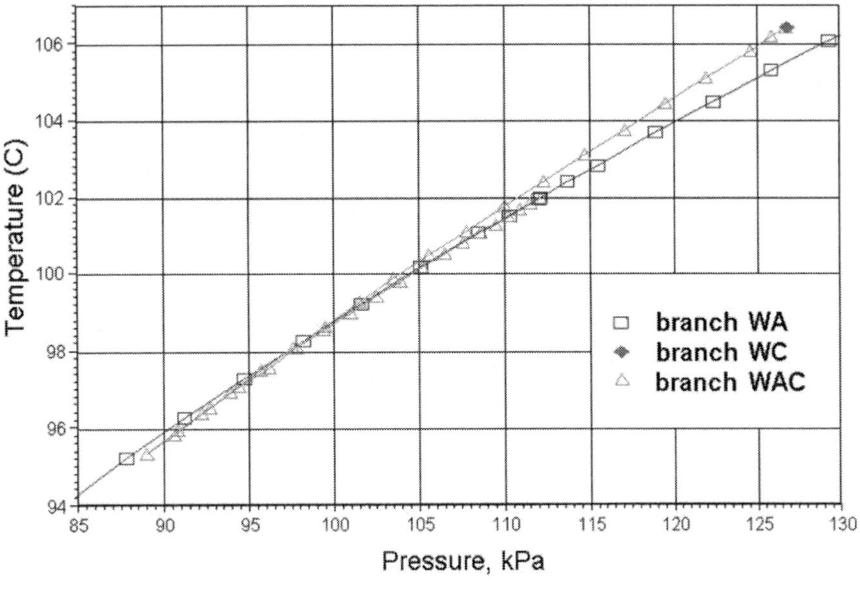

Figure 4. Temperature branches WA, WC and WAC for pressure analysis of azeotropes for ternary mixture W-A-C between 85 and 130 kPa. NRTL-Ideal model.

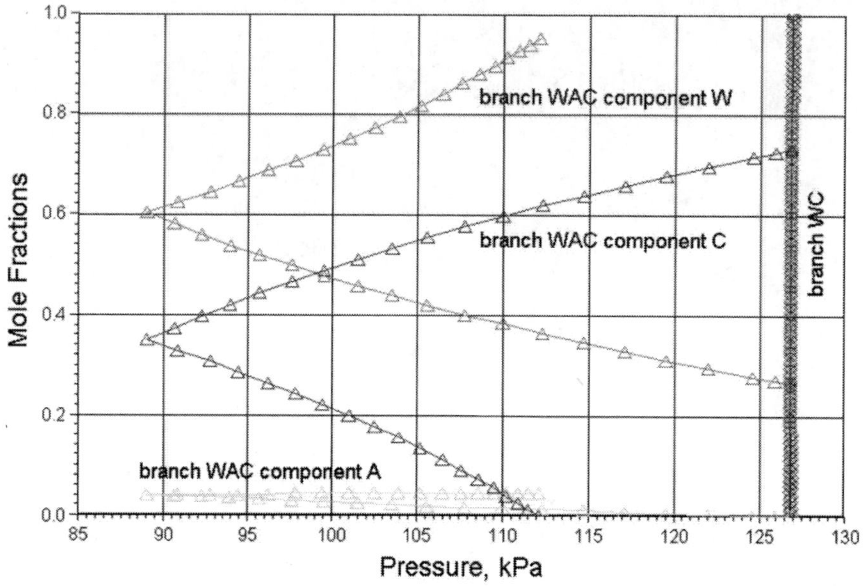

Figure 5. Composition branches WC and WAC for pressure analysis of azeotropes for ternary mixture W-A-C between 85 and 130 kPa. NRTL-Ideal model.

Notice how much more information we have got from APSA compare to DRD diagrams created for a few selected pressures (Figure 2). The DRDs confirm existence of only one azeotrope WA. The binary azeotrope WC exists only in a very narrow range of pressures (see Figure 4 and Figure 5) and is practically visible as a homotopy continuation branch WC only in Figure 3. The ternary azeotrope WAC hasn't been found by azeotrope searching algorithm (see DRD at 100 kPa in Figure 2). It is because the homotopy method used in the algorithm cannot find isolas – a continuation paths not connected to any homotopy branch that starts from a pure component (Fidkowski et al., 1993). Actually model predicts two WAC azeotropes at 100 kPa (see Figure 5). One is a saddle; the other is an unstable node. Because we have missed two azeotropes and their contributions to the Zaharov-Serafimov topological constraint (see e.g. Equation (14) in Wasylkiewicz et al., 1999) cancel each other, the consistency test is fulfilled and there is no warning during creation of DRD at 100 kPa (Distil, 2003).

Troubleshooting of Cyclohexanone – Cyclohexanol – Phenol Distillation

Another problem with simulation of distillation column has been reported for the mixture of cyclohexanone (CC6one), cyclohexanol (CC6ol) and phenol. APSA has been applied to the mixture to verify accuracy of predictions of two VLE models throughout a pressure range of industrial interest. The NRTL model predicts existence of the ternary maximum boiling azeotrope in a wide pressure range (only slightly pressure sensitive) whereas the Wilson model predicts existence of the ternary saddle azeotrope only in a very narrow range of pressures (extremely pressure sensitive). Qualitatively different predictions of the two models are discussed and compared to experimental data. Excellent effectiveness of APSA for validation of VLE models and corresponding interaction parameters is shown later in this paragraph.

For the ternary system in consideration, it is known from measurements that:

- there is a maximum boiling azeotrope in the binary system CC6one – phenol between 8.00 and 101.32 kPa (Gmehling et al., 1994, Part II, p. 1096);
- there is a maximum boiling azeotrope in the binary system CC6ol – phenol between 9.33 and 101.32 kPa (Gmehling et al., 1994, Part II, p. 1097);
- the three-component system does not form a ternary azeotrope at 12.00 kPa (Gmehling et al., 1994, Part II, p. 1710).

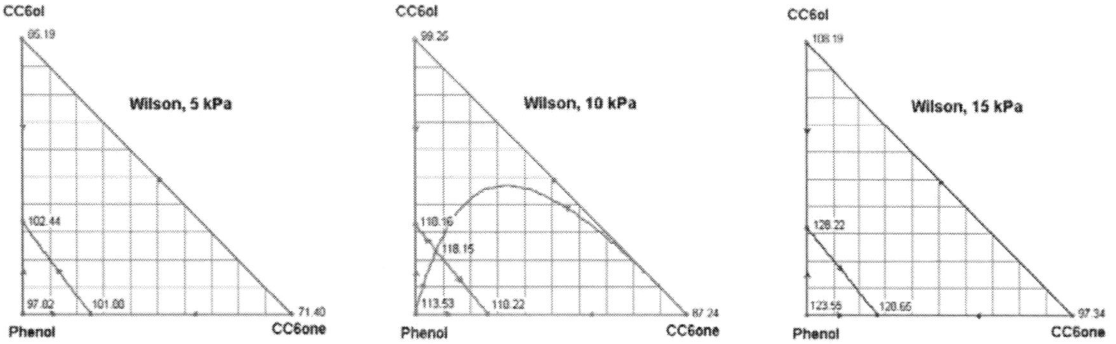

Figure 6. Distillation region diagrams for the mixture of cyclohexanone (CC6one), cyclohexanol (CC6ol) and phenol at 5, 10 and 15 kPa. Based on the Wilson model.

Distillation region diagrams for the mixture are presented in Figure 6 for a few selected pressures. VLE calculations were based on Wilson model. Binary interaction parameters were taken from Dechema, Volume 1, Part 2b, pp. 369, 371 and 403 (Gmehling et al., 1981). Wilson model does not predict existence of the ternary azeotrope at 5 and 15 kPa. There is, however, a very narrow range of pressures, where the ternary azeotrope exists. The temperatures of all azeotropes are so close that any small change in pressure can change their type and the structure of the DRD. At 5 kPa the CC6ol – phenol azeotrope is stable node and the CC6one – phenol azeotrope is saddle whereas at 15 kPa the CC6ol – phenol azeotrope is saddle and the CC6one – phenol azeotrope is stable node.

The homotopy continuation method (Wasylkiewicz et al., 1999) was used to find all azeotropes at particular pressure. The topological consistency was checked using a topological constraint (Zharov and Serafimov, 1975; Fidkowski et al., 1993). All presented diagrams are topologically consistent. If the topological constraint is not fulfilled one or more azeotropes are missing or their properties have been estimated incorrectly. However, fulfillment of the topological constraint does not guarantee that all azeotropes have been found and the system properties have been determined correctly. The topological constraint is necessary but not not sufficient condition for consistency in azeotrope calculations (see e.g. Wasylkiewicz et al., 1999).

Table 2. Results of bifurcation pressure search for the mixture of Cyclohexanol (CC6ol), Cyclohexanon (CC6one) and Phenol between 5 and 15 kPa. Wilson model.

Bifurcation	Pressure, kPa	Bifurcation composition, mole			Temperature, °C
B1	8.26	0.0	0.262584	0.737416	113.54
B2	10.94	0.323241	0.0	0.676759	120.34

The APSA was performed for the mixture between 5 and 15 kPa. Branches of solutions are shown in composition bifurcation diagram (Figure 7). In this pressure range, two bifurcation pressures have been found (see Table 2 for details):

- **B1.** 8.26 kPa, where the ternary CC6one – CC6ol – phenol branch bifurcates from the CC6one – phenol branch; the CC6one – phenol azeotrope changes type from saddle to stable node;
- **B2.** 10.94 kPa, where the ternary CC6one – CC6ol – phenol branch collapses on the CC6ol – phenol branch; the CC6ol – phenol azeotrope changes type from stable node to saddle.

Both binary azeotropes are only a little sensitive to pressure changes. On the other hand, the ternary azeotrope (saddle) is extremely pressure sensitive and exists only in the very narrow pressure range, between the two bifurcation pressures 8.26 and 10.94 kPa. From experiments we know that the three-component system doesn't form the ternary azeotrope at 12.00 kPa (Gmehling et al., 1994, Part II, p. 1710). This agrees well with the Wilson's model predictions. New experiments are needed in the pressure range between the two bifurcation pressures to check the model predictions. It must be also emphasized here that, unlike to stable and unstable azeotropes, it is much more difficult to confirm experimentally the existence of a saddle azeotrope.

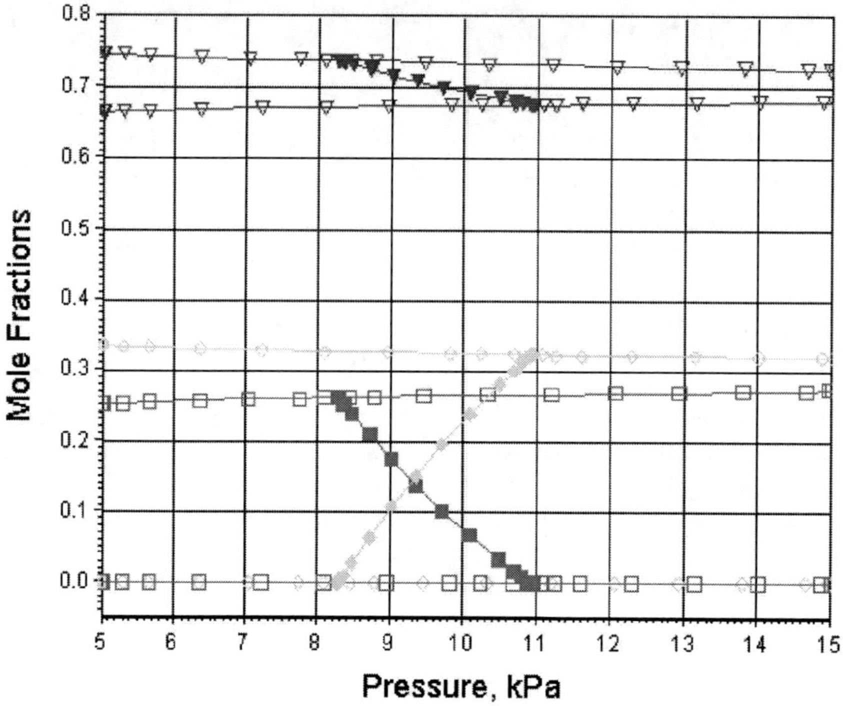

Figure 7. Composition changes of azeotropes with pressure for the mixture of cyclohexanone (squares), cyclohexanol (diamonds) and phenol (triangles) between 5 and 15 kPa. Based on the Wilson model.

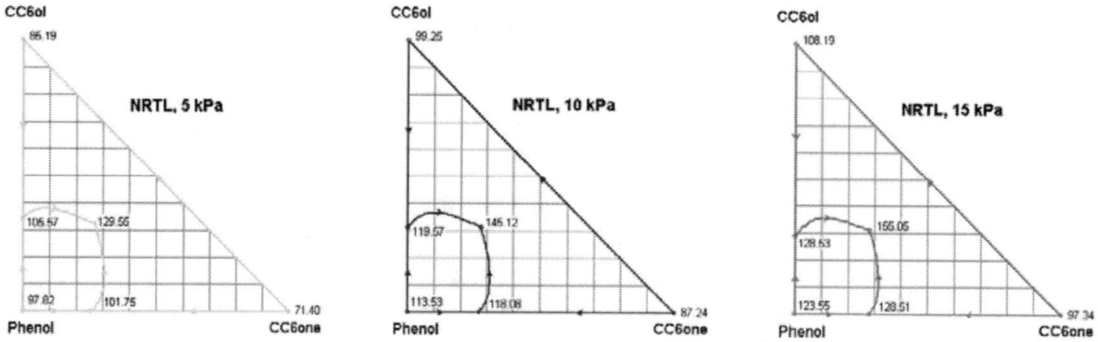

Figure 8. Distillation region diagram for the mixture of cyclohexanone (CC6one), cyclohexanol (CC6ol) and phenol at 5, 10 and 15 kPa. Based on the NRTL model.

Distillation region diagrams for the mixture calculated based on NRTL model are presented in Figure 8 for the same pressures as for the Wilson model above. Binary interaction parameters were taken from Dechema, Volume 1, Part 2b, pp. 369, 371 and 403 (Gmehling et al., 1981). NRTL model predicts existence of the ternary maximum boiling azeotrope for all of these pressures. Even in the wider range of pressures 1 – 100 kPa all three azeotropes (two binary saddles and one ternary maximum boiling) exist and do not change their types.

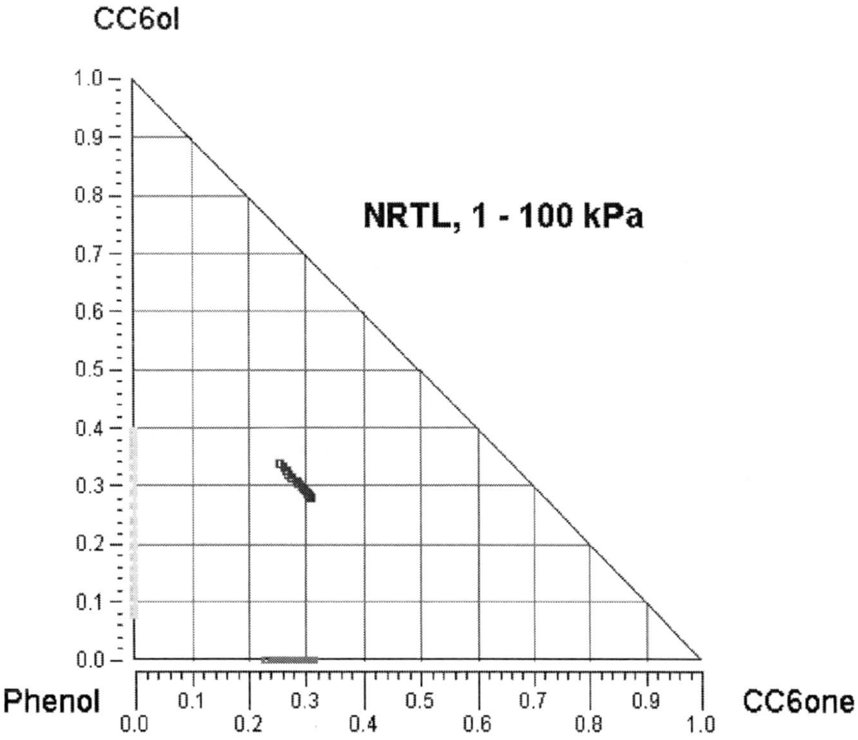

Figure 9. Branches of pressure sensitivity analysis of azeotropes for the mixture of cyclohexanone (CC6one), cyclohexanol (CC6ol) and phenol between 1 and 100 kPa. Based on the NRTL model.

APSA was performed for the mixture in the pressure range 1 – 100 kPa. Branches of solutions predicted by the NRTL model are shown in composition triangle in Figure 9. No bifurcation pressures was found. In this case, the binary CC6ol – phenol azeotrope is the most pressure sensitive azeotrope whereas the ternary maximum boiling CC6one – CC6ol – phenol azeotrope is the least sensitive (Figure 9). This is in contradictory to the predictions of the Wilson model (Figure 7).

The ternary system presented here is a challenging one for both equilibrium and azeotrope calculations. Qualitatively different predictions of the two applied models have been found for the ternary CC6one – CC6ol – phenol azeotrope: extremely pressure sensitive saddle azeotrope in a very narrow range of pressures for Wilson model; moderately pressure sensitive maximum boiling azeotrope in a wide pressure range for NRTL model. At the same time the binary azeotropes are predicted equally well for both models and are consistent with binary experimental data. More experiments for ternary system are necessary especially in the pressure range between the bifurcation pressures (8.26 and 10.94 kPa) to accept or reject Wilson model predictions. Based on the only one available experimental data point for the ternary mixture at 12.00 kPa (Gmehling et al., 1994, Part II, p. 1710) we can conclude that the NRTL model should be corrected (better interaction parameters should be regressed). Generally, if a model is expected to reflect reality, interaction parameters should be regressed for the model using experimental data covering the particular pressure range of interest. In this case, special attention is required for the saturated pressure correlation for pure components (e.g. Antoine equation). Since CC6ol and CC6one are close boiling components, small inaccuracies in prediction of saturated pressure can cause significant changes in the DRD structure.

Conclusions

It is extremely important to select proper thermodynamic model and carefully verify behavior of the mixture for particular set of components before any attempt is made to simulate distillation column. DRDs are extremely useful for this verification because of their visualization capabilities. They are, however, restricted to ternary and quaternary mixtures. There are no such restrictions for azeotrope pressure sensitivity analysis (APSA). Bifurcation pressures can be calculated and temperature and composition bifurcation diagrams can be created for any numbers of components. This provides an excellent tool for both design and troubleshooting of real distillation columns.

The extreme sensitivity of azeotropes to pressure change described in the above two examples has been detected in several distillation systems especially when close boiling components were present. If detected, this extreme sensitivity should trigger more careful examination of the thermodynamic model (applicability, parameters), may be some additional VLE measurements or just draw attention to avoid the particular pressure range both in the design and in the operation of the distillation column.

Changes in product quality and quantity often require operation of distillation columns at different pressures. Even when operating in a near steady-state condition, many columns exhibit significant pressure gradients between the distillate and bottoms. In some cases, this can lead to a switch in topology of distillation regions and cause serious problems in convergence of calculations for steady state or dynamic simulators. APSA provides a dependence of compositions and temperatures of azeotropes with respect to changes in pressure. This information can be of critical importance in the design and troubleshooting of distillation separation systems. APSA can also be used to verify the suitability and accuracy of the VLE models throughout a pressure range if the system is known to have particular sets of azeotropes at particular pressures. The implementation of the method is available in the commercial program Distil (2003).

Notations

a_{ij} parameter of binary interactions between components i and j, cal mol^{-1}
P pressure, kPa
P^s saturation pressure, kPa
T temperature, K (°C in graphs)

Greek letters

α_{ij} NRTL nonrandomness parameter
v_i molar volume of pure component for Wilson equation, m^3 kmol^{-1}

References

Agraval, R., Li, Y. K., Santollani, O., Satyro, M. A., & Vieler, A. (2001). Uncovering the Realities of Simulation, *Chem. Eng. Progress*, Part 1 (May), 42-52; Part 2 (June), 64-72.

Chapman, R.G., & Goodwin, S.P. (1993). A General Algorithm for the Calculation of Azeotropes in Fluid Mixtures, Fluid Phase Equilibria, 85, 55–69.

Distil v. 6.1 (2003). Aspen Technology Inc., http://www.aspentech.com.

Doherty, M.F., & Malone, M.F. (2001). *Conceptual Design of Distillation Systems*. New York, NY: McGraw-Hill.

Fidkowski, Z.T., Malone, M.F., & Doherty, M.F. (1993). Computing Azeotropes in Multicomponent Mixtures, Comput. Chem. Eng., 17, 1141–1155.

Gmehling, J., Onken, U., & Arlt, W. (1981). *Vapor-Liquid Equilibrium Data Collection. DECHEMA Chemistry Data Series,* Vol. I, Frankfurt: DECHEMA.

Gmehling, J., Menke, J., Fischer, K., & Krafczyk, J. (1994). *Azeotropic Data*, Weinheim: VCH Verlagsgesellschaft mbH.

Harding, S.T., Maranas, C.D., McDonald, C.M., & Floudas, C.A. (1997)., Locating All Homogeneous Azeotropes in Multicomponent Mixtures, Ind. Eng. Chem. Res., 36, 160–178.

Horsley, L. H. (1947). Graphical method for predicting azeotropism and effect of pressure on azeotropic constants. Analyt. Chem., 19, 603–609.

Knapp, J. P. (1991). *Exploiting pressure effects in the distillation of homogeneous azeotropic mixtures.* Ph.D. Thesis, University of Massachusetts, Amherst.

Maier, R.W., Brennecke, J.F., & Stadtherr, M.A. (1998). Reliable Computation of Homogeneous Azeotropes, AIChE J., 44, 1745–1755.

Malesinski, W. (1965). *Azeotropy and Other Theoretical Problems of Vapor-Liquid Equilibrium*, New York: Interscience.

Perry, R. H., Green, D. W., & Maloney, J. O. (1997). *Perry's Chemical Engineers' Handbook*, 7th edition, pp. 13-59 – 13-65, New York : McGrae-Hill.

Roscoe, H. E., & Ditmar, W. (1859). On the absorption of the hydrochloric acid and ammonia in water. J. Chem. Soc., 12, 128–151.

Roscoe, H. E. (1860). On the composition of the aqueous acids of constant boiling point. J. Chem. Soc., 13, 146–164.

Roscoe, H. E. (1862). On the composition of the aqueous acids of constant boiling point – second communication. J. Chem. Soc., 15, 270–276.

Villiers, W.E., French, R.N. & Koplos, G.J. (2002), Navigate Phase Equilibria via Residue Curve Maps, *Chem. Eng. Progress*, November, 66-71.

Wasylkiewicz, S.K., Sridhar, L.N., Doherty, M.F., & Malone, M.F. (1996). Global Stability Analysis and Calculation of Liquid-Liquid Equilibrium in Multicomponent Mixtures, Ind. Eng. Chem. Res., 35, 1395–1408.

Wasylkiewicz, S.K., Malone, M.F., & Doherty, M.F. (1999). Computing All Homogeneous and Heterogeneous Azeotropes in Multicomponent Mixtures, Ind. Eng. Chem. Res., 38, 4901–4912.

Wasylkiewicz, S.K., Kobylka, L.C., & Castillo, F.J.L. (2000). Pressure Sensitivity Analysis of Azeotropes in Synthesis of Distillation Column Sequences, Hungarian Journal of Industrial Chemistry, 28, 41–45.

Wasylkiewicz, S.K., & Shethna, H.K. (2002). VLE Data Estimation for Synthesis of Separation Systems for Azeotropic Mixtures, AIChE Spring National Meeting, paper No. 4e, New Orleans, LA, March 2002.

Wasylkiewicz, S.K., Kobylka, L.C., & Castillo, F.J.L. (2003). Pressure Sensitivity Analysis of Azeotropes, Ind. Eng. Chem. Res., 42, 207–213.

Wasylkiewicz, S. K. (2004a). Diagrams troubleshoot acetic acid dehydration simulation, contribution to the book "Distillation Troubleshooting" edited by Henry Z. Kister, John Wiley & Sons, Inc., in press.

Wasylkiewicz, S.K. (2004b). Modeling ternary mixture using binary interaction parameters, contribution to the book "Distillation Troubleshooting" edited by Henry Z. Kister, John Wiley & Sons, Inc., in press.

Zawisza, A. C. (1961). Change of azeotropic parameters with pressure, Bull. Acad. Polon. Sci. Ser. Sci. Chem., 9, 147–150.

Zharov, W. T., & Serafimov, L. A. (1975). *Physicochemical Fundamentals of Distillation and Rectification*, Leningrad: Khimiya.

APPENDIX: Thermodynamic data for CC6one – CC6ol – phenol system

Vapor pressure expression (Distil, 2003):

$$\ln(P^s) = A + \frac{B}{T} + D\ln(T) + ET^2,$$

P^s in kPa, T in K.

Component	A	B	D	E	v_i
CC6one	81.0891	-8341.16	-9.5685	5.30989e-6	0.10418
CC6ol	195.740	-14920.5	-26.2134	1.30575e-5	0.10343
phenol	128.846	-11676.5	-16.3690	7.84630e-6	0.083138

Wilson binary interaction parameters, a_{ij} [cal/mol]:

	CC6one	CC6ol	phenol
CC6one	0	1.372799	-602.4887
CC6ol	98.63539	0	-377.2987
phenol	-814.3267	-815.3192	0

NRTL binary interaction parameters, a_{ij} [cal/mol]:

	CC6one	CC6ol	phenol
CC6one	0	129.0207	-625.6519
CC6ol	-29.4801	0	-59180.55
phenol	-873.0073	65670.6	0

NRTL nonrandomness parameters, α_{ij}:

	CC6one	CC6ol	phenol
CC6one	0	0.3013	0.3100
CC6ol	0.3013	0	0.0016
phenol	0.3100	0.0016	0

Binary interaction parameters were taken from Gmehling et al. (1981) Vol. 1, Part 2b, pp. 369, 371 and 403.

81c Using Evolutionary Operation (EVOP) to Optimize Distillation Operations

Scott Moffatt, Valerie Monical

Normal technical effort to improve the performance of a distillation process will typically yield improvements in time. Targeting these efforts using a technique called evolutionary operation (EVOP) can yield accelerated imrovements. EVOP is most useful in continuous processes where the optimum operation is a moving target due to upstream changes or changing product requirements. EVOP uses intentional but slight process changes designed to improve the signal-to-noise ratio and allow discovery of effects that are normally too small to see. The advantages of the EVOP technique include optimization without going outside the normal operating conditions of the process, continuous use of the process to produce information on how to best produce the final product, no issues of scale-up, and no complex system modeling. This paper will illustrate through case studies the use of theEVOP technique to optimize product impurity levels and utility costs using distillation system parameters such as reflux rate, recycle rates, and temperature set points

81d High Mass Transfer Efficiency with Integrated Heat Transfer

Anna Lee Tonkovich, Laura Silva, Ravi Arora, Tom Hickey

Velocys, in partnership with ABB Lummus Global, is developing a potential breakthrough in technology for ethane-ethylene fractionation using microchannel processing in a project sponsored by the U.S. Department of Energy. The process is expected to provide significant benefits in energy utilization and capital expenditure compared to conventional distillation technology.

Velocys® microchannel technology enhances mass transfer by contacting thin vapor and liquid films in microchannels, dramatically reducing the Height-to-an-Equivalent Theoretical Plate (HETP) for distillation. Temperature profiles are tailored and adjusted along the length of process microchannels through adjacent heat exchange microchannels integral to the separation hardware, rather than in separate heat exchangers. Enhancements in both mass transfer and heat integration can enable process flowsheet options which have been previously commercially unattractive. The microchannel technology is scaled to commercial plant capacities by increasing the numbers of microchannels conducting parallel processing. Techniques for design and fabrication of scaled-up microchannel processing plants have been developed by Velocys for large-scale micro reactor applications.

This presentation will cover the technical approach, benefits, and recent accomplishments in the development of the Velocys distillation process.

81e From Molecular Structure to Distillation Design: The SPEAD Model

J Richard Elliott

Discontinuous molecular dynamics simulation (DMD) and thermodynamic perturbation theory (TPT) have been used to study thermodynamic and dynamic (transport) properties for a wide range of compounds.[1, 2] The fundamental basis of the method relies on a stepwise characterization of disperse interactions[3] and Wertheim potentials for hydrogen bonding.[4] Transferable step potentials are characterized based on vapor pressure and liquid density from available experimental data. An interaction site diameter and the depths of the inner and outer wells of the potential characterize the transferable multi-step potentials. Intermediate wells are interpolated reducing the number of adjustable parameters to three per site type. The combined use of step potentials for equilibrium and dynamic properties is referred to as the SPEAD model.

Overall, the vapor pressures are correlated to within ~15% error and liquid densities to 4% error with reduced temperatures generally extending to 0.45. Combined with simulation results for mixture entropy and enthalpy, a complete model for distillation design is made feasible. This model has been implemented into the ChemCAD process simulator. The complete methodology from molecular structure to process simulation will be demonstrated.

1. Unlu, O., Gray, N., Gerek, Z. N., Elliott, J. R. 2004. Ind. Eng. Chem. Res. 43: 1788

2. Gray, N. H., Gerek, Z. N., Elliott, J. R. 2005. Fluid Phase Eq. submitted

3. Barker, J. A., Henderson, D. 1967. J. Chem. Phys. 47: 2856

4. Wertheim, M. S. 1984. J. Stat. Phys. 35: 19

81f Retrofit Distillation Design Using Shortcut Models for Simulation and Hydraulic Analysis

Megan Jobson, Walter Castillo

Objectives of retrofit design of existing distillation columns include increasing throughput, accommodating a change in feedstock composition and achieving a new set of product specifications. The retrofit design options that engineers must evaluate in order to meet these objectives include replacing the column internals with valve trays, structured packing or high performance trays, modifying the operating conditions (e.g. feed condition, feed location, operating pressure), introducing intermediate heating or cooling and installing a prefractionator. Current design practice involves trial and error, based on engineering experience, and rigorous simulation to explore the many options, together with hydraulic analysis. This work proposes short-cut models for column simulation and hydraulic analysis to allow promising design options to be identified simply and quickly. The hydraulic analysis assumes that entrainment flooding limits flows in the column.

A method for hydraulic analysis of a distillation column using rigorous simulation results was proposed by Liu and Jobson (2004). The fractional utilisation of area (FUA) on each stage in the column was proposed to identify the extent to which available area in the column is utilised. The FUA is defined as the ratio of the area required for vapour flow on a stage to the area available for vapour flow where entrainment flooding is assumed to limit operation. Analysis of the FUA curve provides quantitative and qualitative guidelines for retrofit design. However, the FUA curve requires rigorous simulation results for several design options. This work develops a short-cut approach for hydraulic analysis and evaluation of design options, which uses the results of short-cut column simulations, based on the Fenske-Underwood-Gilliland (FUG) method.

The FUG method provides information about flow rates and compositions of the column products, but not about composition profiles within the column. In this work, a boundary method is employed in order to calculate the composition profile within an existing column, together with an energy balance at the feed stage. The Fenske equation (King, 1971) is used to estimate the distribution of the non-key components in the product streams. The composition profile provides the necessary information to estimate the fluid properties involved in the calculation of the column diameter, such as density, mean molecular weights and surface tension. The estimation of column diameter uses established hydraulic correlations, and the boundary method is applied by using only short-cut simulation results.

From this FUA curve, the maximum achievable flows in the column can be determined and the stages creating a bottleneck in the column can be identified. Design options aiming to overcome this bottleneck can be evaluated using repeated short-cut simulation; the benefit, e.g. in terms of throughput, of changing the thermal condition of the feed, the operating pressure, the number of theoretical stages, or the type of internals may be quantified. The short-cut method for hydraulic analysis forms the basis for a new retrofit design tool applicable to simple columns separating mixtures for which the assumption of constant relatively volatility is valid.

The results of the short-cut hydraulic analysis are compared to those obtained by rigorous simulation for various systems to demonstrate the validity of the short-cut hydraulic model. The new approach is applied to a range of problems to evaluate and screen retrofit design options.

81g A New Turbulent Model for Computational Mass Transfer and Its Application to Commercial-Scale Distillation Column

B T Liu, Chunjiang Liu, K T Yu, X G Yuan, Z.M. Sun

The key problems of applying the computational mass transfer to the distillation column are the enclosure of the corresponding differential equation of mass transfer and the formulation of the related package of computational fluid dynamics (CFD). Since many researchers have investigated the CFD of a distillation tray, the remaining difficulty is to solve the enclosure problem. In this paper, the authors propose a methodology of enclosure, designated as model. The combination of the proposed model equations, the corresponding differential mass transfer equation and the relevant set of CFD equations are solved simultaneously, so that the velocity and concentration profiles as well as the Murphrey efficiency of each tray of the whole distillation column can be predicted at the same time. The main advantages of the new method of simulation are as follows. Firstly, the unknown turbulence mass transfer diffusivity, which is hardly found from the literatures and should be determined empirically in the traditional approaches, can be eliminated. Secondly, as the flow/concentration profiles and the efficiency of each tray could be reliably predicted and previewed in the early stage of design, the imperfections of velocity and concentration distribution of a column could be easily and timely detected, which is beneficial to the designer to make necessary changes to achieve better efficiency. The applicability of present method is illustrated by simulating a commercial scale distillation column with 8 sieve trays of 1.2 meters diameter. The results of comparison between the predictions and the experimental data are satisfactorily and promising. This work also demonstrates the significance that applying the computational mass transfer and CFD methods to the simulation of chemical equipment is the role of approaching engineering problems with scientific fundamentals and providing basis for in-depth analysis of the process efficiency, whereas the traditional method is unable to accomplish.

81h Process and Mechanical Design of Methanol Distillation System

Jinsheng Sun, Tao Wang, Xingang Li, Yufeng Tian

A very important material in organic chemical industry, methanol has been used in manufacture of plastic material, synthetic fiber, synthetic rubber, dyestuff, covering material, spice, medical field, pesticide and etc widely. Also, as a fine solvent, it can not only solve many kinds of organic compounds but also be a sound choice for green energy---as substitute of gasoline in many countries. With the quick development of the national economy, the demand (or potential demand) of methanol has sharply increased, which accelerates the development of the methanol industry and increment of the methanol output. Presently, methanol distillation technology has been developed to be large-scale, low energy consuming and more environmental. All the main patent dealers in this industry of the world have the equipments designed for 5000~7500t/d. But China has hardly no running units with capability of 200Kt/a of high quality (such as O-M-232E US). Recently, National Engineering Research Center for Distillation Technology at Tianjin University (NERCDT) has done a lot of work on market and technical research.

Many materials, for instance, natural gas, gas oil, fuel oil, coal, etc., have been used to synthesize methanol. Except for the difference in catalysts and reaction mechanics, the subsequent distillation process and equipments are similar, falling into three main assortments: 2-column process, 3-column process and 3+1-column process. 2-column process is composed of light ends column and methanol refining column or main column which is of found in factories under 100Kt/a, offering simple equipments and low construction investment. 3-column process is built up by light ends column, high pressure methanol refining column and low pressure methanol refining, with the mixture of the distillates from the 2 methanol refining columns the methanol product. The characteristic of this process is the heat exchange between the top of the high pressure methanol refining column and the bottom of the low pressure methanol refining column which saves the energy and the consumption of cooling water. If a methanol recovery column, or stripper, or water column involves after the low pressure methanol refining column to improve the quality of the wastewater, the process will turn into 3+1-column process.

NERCDT completed the simulation of all the above mentioned flowsheets, especially studying the heat exchange between the 2 methanol refining columns. As a summary of our work, the high pressure process and the low pressure process for 3+1-column process was put forward.

A high pressure one takes side draw from exhaustion section of the low pressure methanol refining column the feed of methanol recovery tower (or stripper) on the basis of 3-column process. Most of the purified wastewater drains from the low pressure methanol refining column. The process requires much higher bottom temperature in order to ripe off methanol and other light organic impurities in wastewater. Because the thermal resource of reboiler for this column is the overhead methanol vapor in the high pressure methanol refining column, to maintain normal operation, the operation pressure of the high pressure methanol refining column must be high enough to offer the temperature gradient in heat exchange (condenser/reboiler) .

The low pressure3+1-column process is that the low pressure methanol refining column keeps certain content of methanol in the liquid phrase at the bottom, with the exhaustion the feed of recovery tower (or stripper). Because the boiling point of wastewater in this column is lower than that of the upper case, the high pressure methanol refining column can operate at lower pressure to give enough heat.. The disadvantage of this process is the lower methanol concentration of the feed to recovery column (or stripper) and the larger flow rate increasing the separation difficulty of methanol recovery column (or stripper), which accordingly make bigger size of the column.

For hydraulic design, structure packing is recommended for the whole light ends column, the distillation section of low pressure methanol refining column and part of the exhaustion section between the feed and side draw; valve tray for whole high pressure methanol refining column and the section from side draw to bottom in low pressure methanol refining column to facilitate the arrangement optional side draw location. The methanol recovery column, or stripper, adopts similar structure as the low pressure methanol refining column. At the pressure in high pressure methanol refining column, it makes almost no difference using tray or structure packing, the whole pressure drop takes small ratio to the operating pressure. That is, using the two kinds of column structure in this equipment will get nearly the same bottom temperature, with no impact on outside heat exchange network. That's how an economical low energy consumption design comes out.

In the specific design of the tower internals ,using patent technique of Tianjin University, such as, the large-diameter gauze pie strengthening technique, one-stage large-diameter trough liquid distributor, new chimney tray liquid distributor, guiding trapeze valve tray will get higher distillation efficiency.

REMOVING PACKINGS FROM HEAT TRANSFER SECTIONS OF VACUUM TOWERS

Silvia Waintraub[1]
Glaucia Alves da Silva Torres[2]
Ana Lídia Wild Serpa[3]
Romeu Luis Waschburger[3]
André Gonçalves Oliveira[4]

[1]PETROBRAS/RESEARCH CENTER
[2]PETROBRAS/HEADQUARTERS
[3]PETROBRAS/REFAP S.A
[4]ESSS – Engineering Simulation and Scientific Software

Rio de Janeiro
Brazil

Prepared for Presentation at the AIChE 2005 Spring National Meeting,
Atlanta, April 10 - 14

March/2005

Unpublished

Abstract

Recent papers published by Petrobras related to empty spray sections in vacuum towers showed their viability and effectiveness in heat transfer applications. In this work, the subsequent studies related to this subject will be presented, such as:

- Construction of a pilot plant vacuum tower;
- Use of a small refinery vacuum tower (REMAN) for tests and new implementations as benchmark, and;
- Implementation in the REFAP refinery revamp.

Introduction

Previous studies (Cai & Kunesh, 1999, Kunesh, 1993, Ognisty, 1990, Ognisty & Kao, 1988) have shown that spray nozzles are very efficient direct contact heat transfer devices and that virtually the entire heat transfer occurs in a short distance from the nozzle outlet.

In vacuum towers the lower the pressure drop the deeper is the crude cut point. This is why the majority of the refinery vacuum columns have packing beds. When the packing bed is removed remaining the spray distributor, the region is called an empty spray section.

The use of empty spray sections in heat transfer regions of vacuum towers allows a high deep cut operation, due to the reduction of the pressure drop along the column, resulting in a lower flash zone pressure, increasing gasoil production, and decreasing vacuum residue yield.

In previous papers (Barbabela and alii, 2004, Waintraub and alii, 2003) the results of plant tests, and CFD analysis in empty spray sections of Petrobras vacuum towers were presented. A summary of these studies is shown below:

- The use of two levels of distributors is recommended for improving the spray cones stability, reducing the effective height of the heat transfer section and increasing the capacity;
- The heat transfer for subcooled liquid reflux occurs at the bottom of the region as at the upper section temperature is constant;
- The heat transfer in empty spray sections depends on both liquid and vapor rates;
- When increasing the liquid rate, the heat transfer is increased requiring a lower height;
- When increasing the vapor rate, the height necessary to a given exchange is higher;
- There is a maximum value of the vapor rate (C-factor) above which the stability of the cone shape of the spray is lost.

Based on the knowledge of the maximum C-factor value up to which empty spray sections still work, a procedure to estimate Cs or F-factor of any tower region was developed based only on operational and laboratory data. All Petrobras vacuum towers were diagnosed using this methodology, and two were selected to have their packing removed.

In the present paper we are going to describe these two case studies with their new tower configurations, results of feed inlet distributor CFD modeling, and also the description of a new pilot plant built to study different configurations of vacuum towers.

First Case Study: REMAN refinery

REMAN is Petrobras smallest refinery with two crude distillation units of 2,000m^3/d (U-2110) and 5,000m^3/d (U-2111) capacity.

U-2110 Vacuum Tower (F-2201) was chosen to be Petrobras deep cut benchmark and a unit for testing new configurations and internals. Its process sketch is shown in Figure 1.

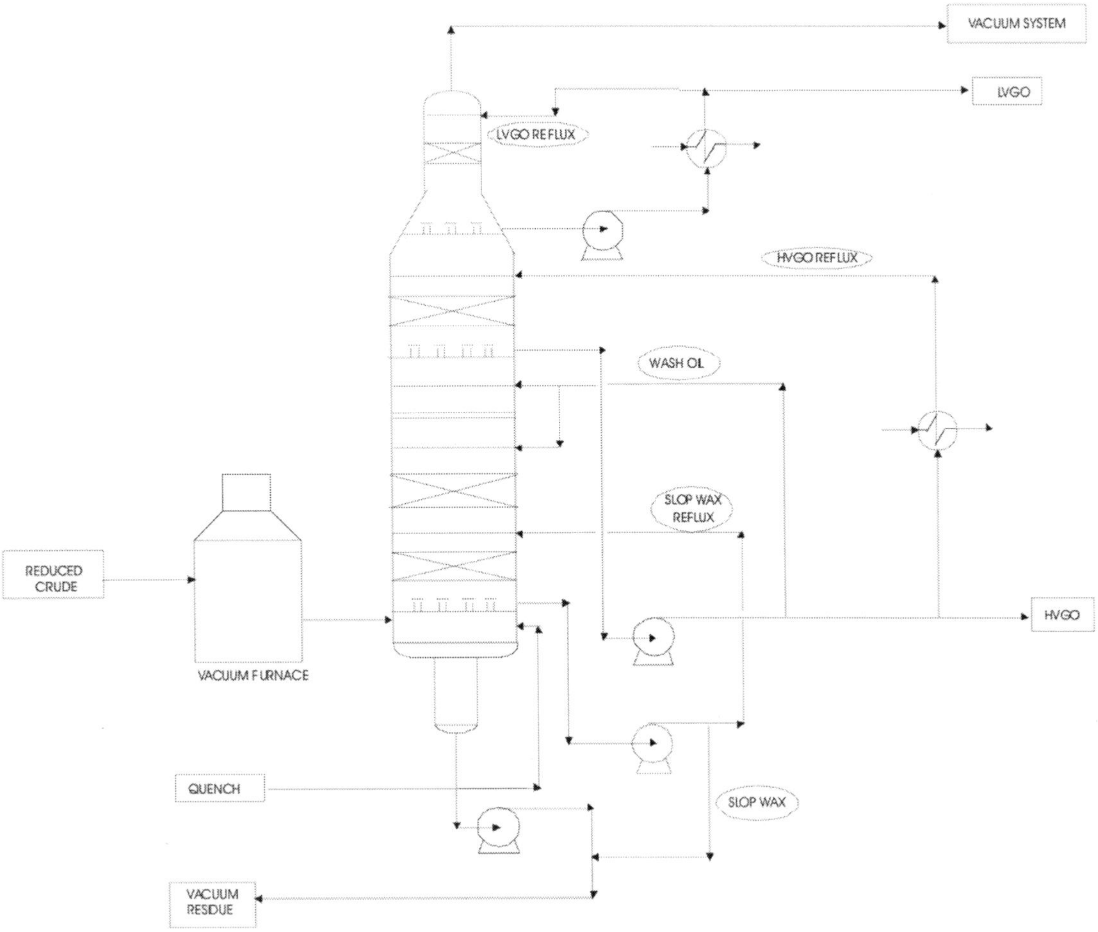

Figure 1 REMAN Vacuum Tower Sketch

A run test was performed in order to allow the identification of the unit limitations related to processing heavy oil and operating with high coil outlet temperature.

Four equally spaced radial skin thermocouples were installed in thirteen levels as shown in Figure 2. These fifty two temperature measurements are sent directly to a computer system, allowing their on line reading.

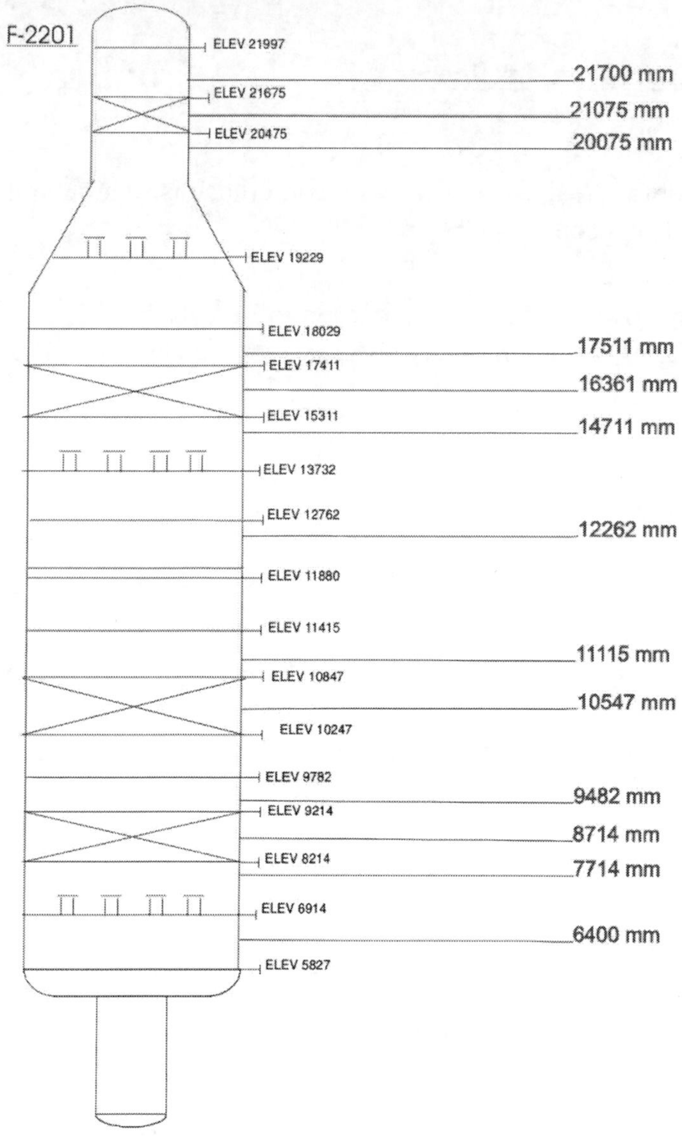

Figure 2 Radial Temperature Levels

A picture of two of these thermocouples is presented in Figure 3. Two different levels are shown and it is important to notice that they are vertically aligned along the tower.

Figure 3 Radial Thermocouples

The major limitation detected during the test was the vacuum furnace. Another observed point was the very low LVGO production due to the use of a very high HVGO reflux rate. In order to increase the C-factor in this region to allow the study of empty spray section, the HVGO reflux rate was drastically reduced.

The results of this diagnose test run were collected, analyzed, and stored to be considered the base data, before future modifications that will be done step by step. The first new configuration proposed was the removal of the LVGO packing, creating an empty spray section, and the installation of two new LVGO spray distributors in two different levels. These distributors were designed to handle 50-100% of the total flow rate, and the allowable pressure drop of the lower one being greater than the upper one. The new configuration is shown in Figure 4.

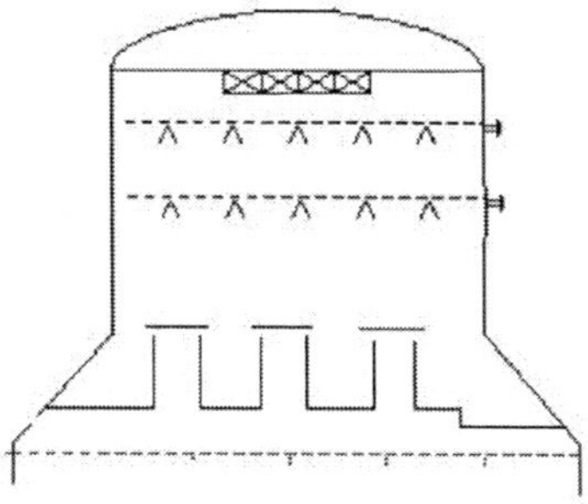

Figure 4 REMAN Vacuum Tower New Configuration

Second Case Study: REFAP refinery

REFAP refinery has a 10,000m³/d distillation unit. This unit was originally designed in 1968 to process 7,500m³/d of crude, and its vacuum distillation started up in 1970 with a feed rate of 3,700m³/d of reduced crude.

In 1991 the topping and vacuum units were revamped to 12,000m³/d and 5,000m³/d, respectively. In 1998 occurred the last revamp in which the LVGO x HVGO fractionation bed was removed and an empty spray section was created. The current vacuum tower configuration is shown in Figure 5a.

Due to the increase of naphthenic heavy crude oils processing, the vacuum tower will be revamped to 5,600m³/d in 2006. The use of empty spray has been proposed for the LVGO and the dirty slop wax pumparounds, as presented in Figure 5b. The C-factors in these sections were obtained using the Petrobras process simulator PETROX, and are 0.05 and 0.30 ft/s, respectively.

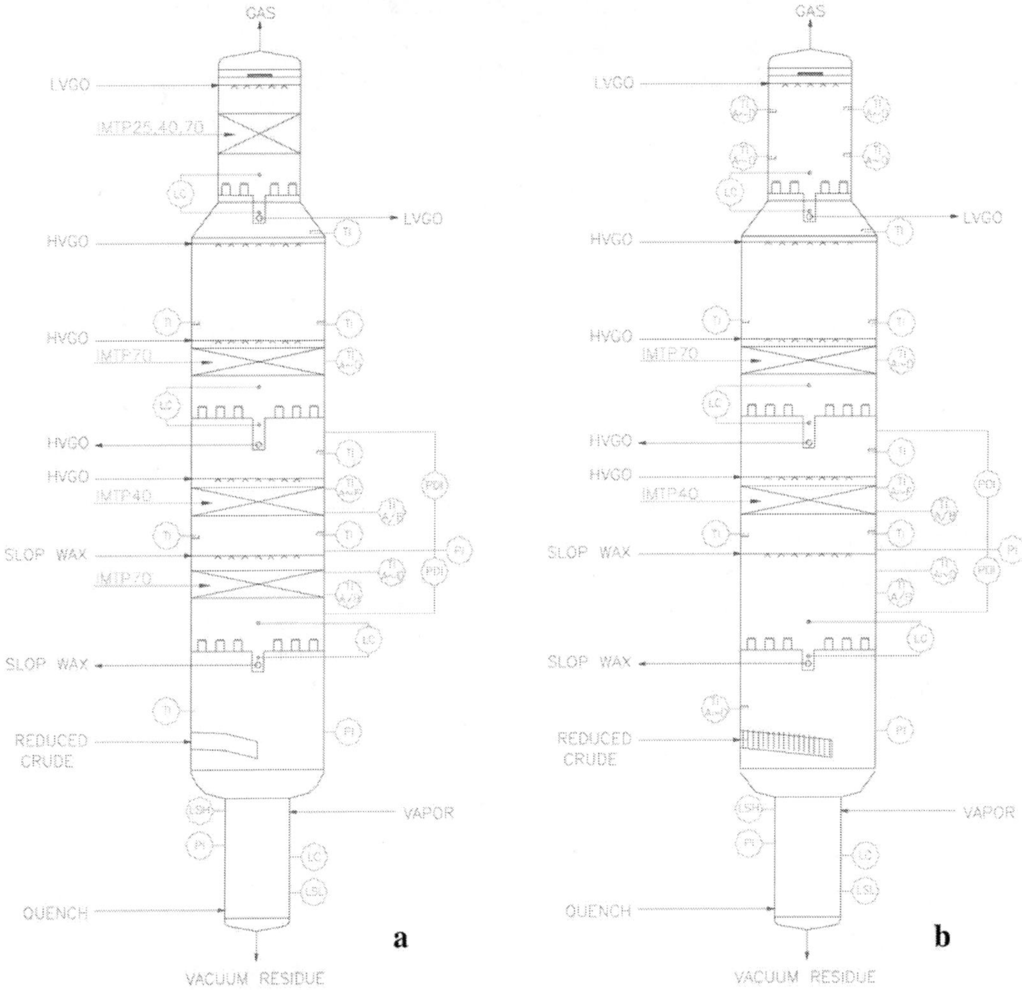

Figure 5 REFAP Vacuum Tower (a) before and (b) after Revamp

The use of empty spray in the LVGO pumparound section requires the substitution of the spray distributor. The number of nozzles is being increased from nineteen to thirty in order to guarantee good distribution. Sixteen thermocouples (eight for the liquid phase and eight for the vapor phase) distributed in two elevations are also being installed to monitor the process.

Due to the use of empty spray in the section just above the flash zone, it is very important to guarantee the vapor distribution from the feed inlet distributor. In order to have a good performance of the dirty wash empty section, it is mandatory to have a good vapor initial velocity distribution to avoid local peaks as well as entrainment. The CFD analysis of this region is described in the next section.

CFD model to REFAP refinery

A computational fluid dynamic simulation was performed by the Distillation CFD Group at Petrobras Research and Development Center (CENPES), using the existing conventional 90° vapor horn, which pictures are shown in Figure 6.

Figure 6 Existing Vapor Horn

CFD results showed that vapor distribution is not uniform, as presented in Figure 7. The main cause is the very high velocities found in the flash zone, mainly at the north, northeast, and northwest regions.

Figure 7 Flash Zone and above Collector Vapor Vertical Velocity Distribution

This non-uniform behavior is confirmed by refinery data that shows an uneven vapor temperature distribution in the tower perimeter, possibly caused by the initial vapor velocity field. These operational data can be seen in Figure 8 and Figure 9.

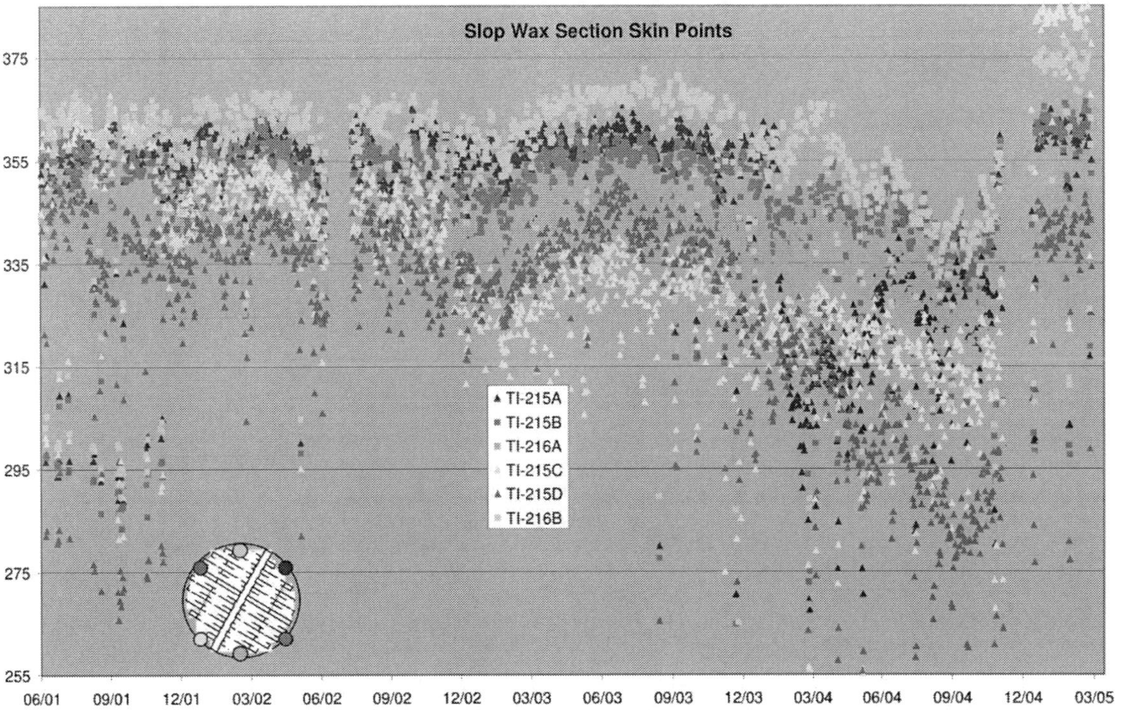

Figure 8 Slop Wax Section Operational Data

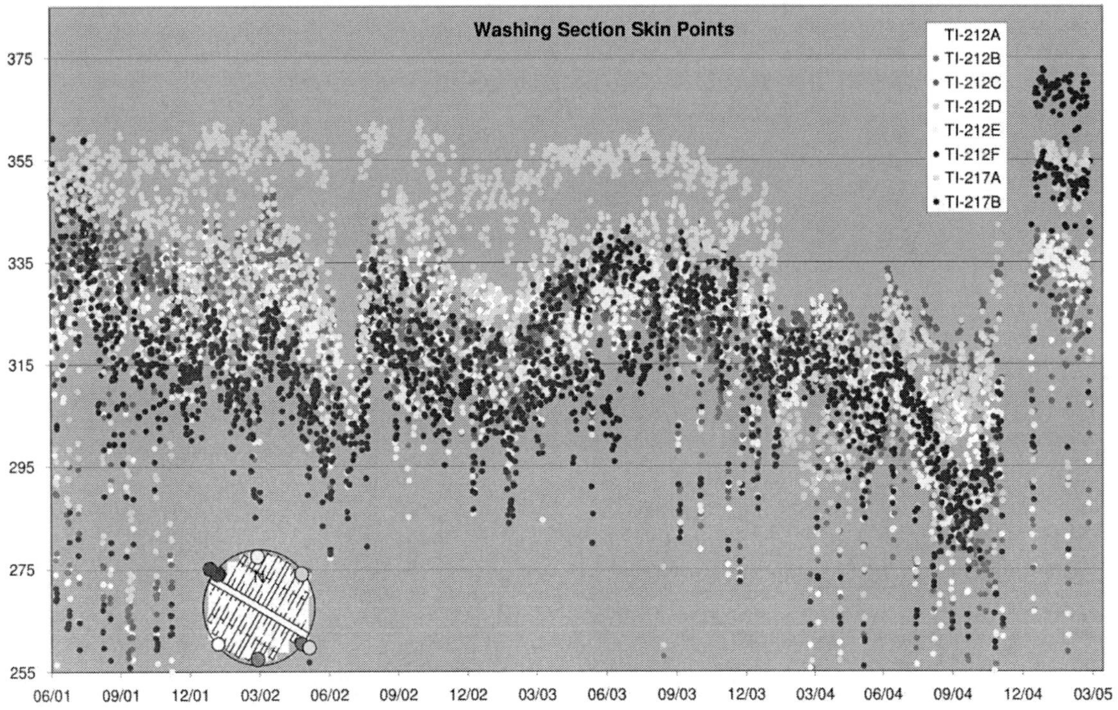

Figure 9 Slop Wax Section Operational Data

This feed inlet device configuration also causes a high liquid entrainment to the region above. This result is shown in Figure 10.

Figure 10 Liquid Wall Volume Fraction Field

Other important variable is the C-factor (Cs). There is an average Cs that is calculated using its correlation, and there are many local C-factors. 3-D plots of the local Cs below and above the slop wax collector pan are shown in Figure 11. High Cs values on the north region, and a better uniformity in the values above the slop wax collector can be noticed. The plane crossing the Cs surface was used just to obtain a better visualization. Below the plane the surface express negatives C-factors values due to the recirculation zone below the collector.

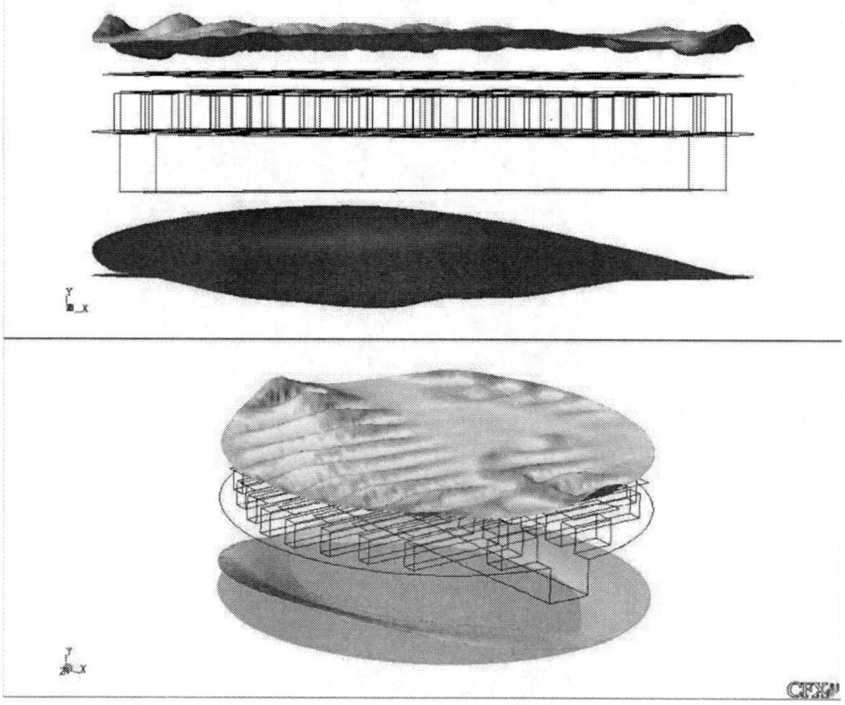

Figure 11 Cs 3D Plot Below and Above the Slop Wax Collector

Figure 12 shows Cs plots of two perpendicular lines below the slop wax collector crossing the tower. It can be noticed that the Cs variation range is about six times negative up to twelve times positive in relation to the average value.

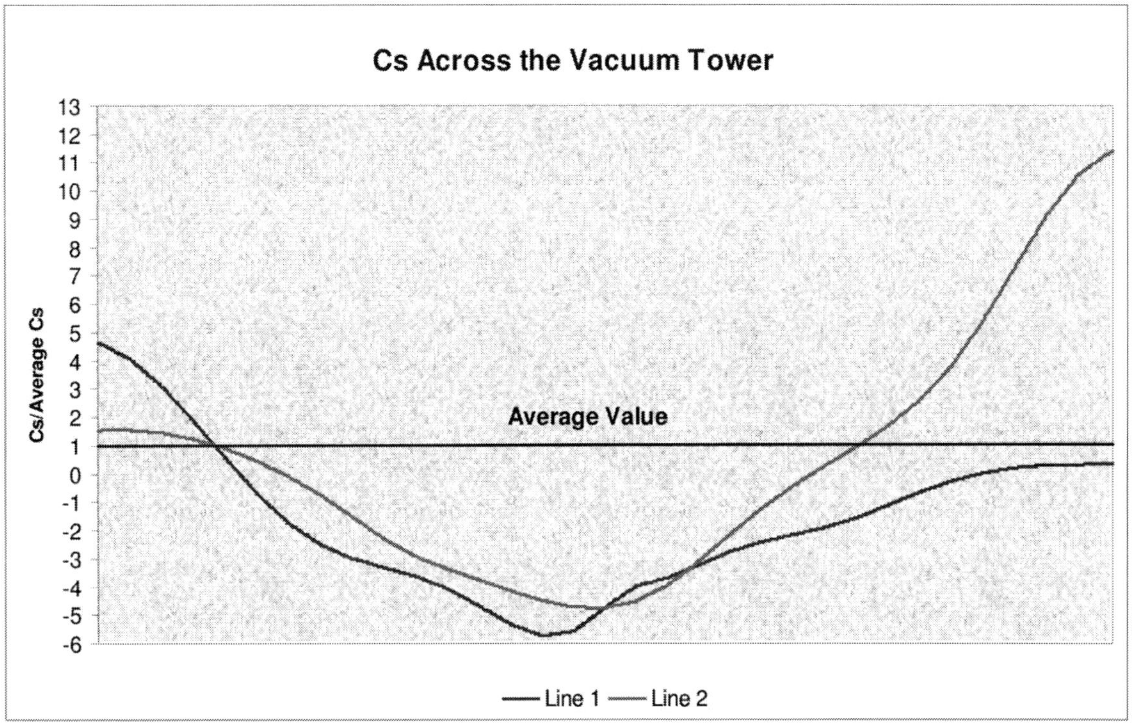

Figure 12 C-Factor and Average C-Factor below the Slop Wax Collector

From the CFD results it was concluded that the vapor velocity was too high and unevenly distributed, causing liquid entrainment to the section above.

Although the global C-factor was not very high, extremely high local values were found, which indicates that the use of empty spray section is not practicable.

As a result of this study, a new modern vapor horn was designed by manufactory company.

Construction of a Vacuum Tower Pilot Plant

Due to the processing of increasingly naphthenic heavy Brazilian crude oils, it is very important to optimize the operation of all vacuum towers, studying different configurations. With this purpose Petrobras decided to build a vacuum pilot plant, which will also be used to test different alloys, process simulation validation, and determination of cracked gases formation correlations, among others.

This tower has a diameter of 280mm, it is composed by separated modules, four side draws, fractionation and pumparound beds between all product draws. These modules can be relocated or removed, giving an excellent configuration flexibility. The major modules are: fractionation, heat exchange, draw-off collectors, distributors, and feed inlet. The tower can operate either as a dry, a dump or a wet vacuum. Process sketch and a preliminary PFD are shown, respectively, in Figure 13 and Figure 14, and pictures of the place where it will be installed are presented in Figure 15 and Figure 16.

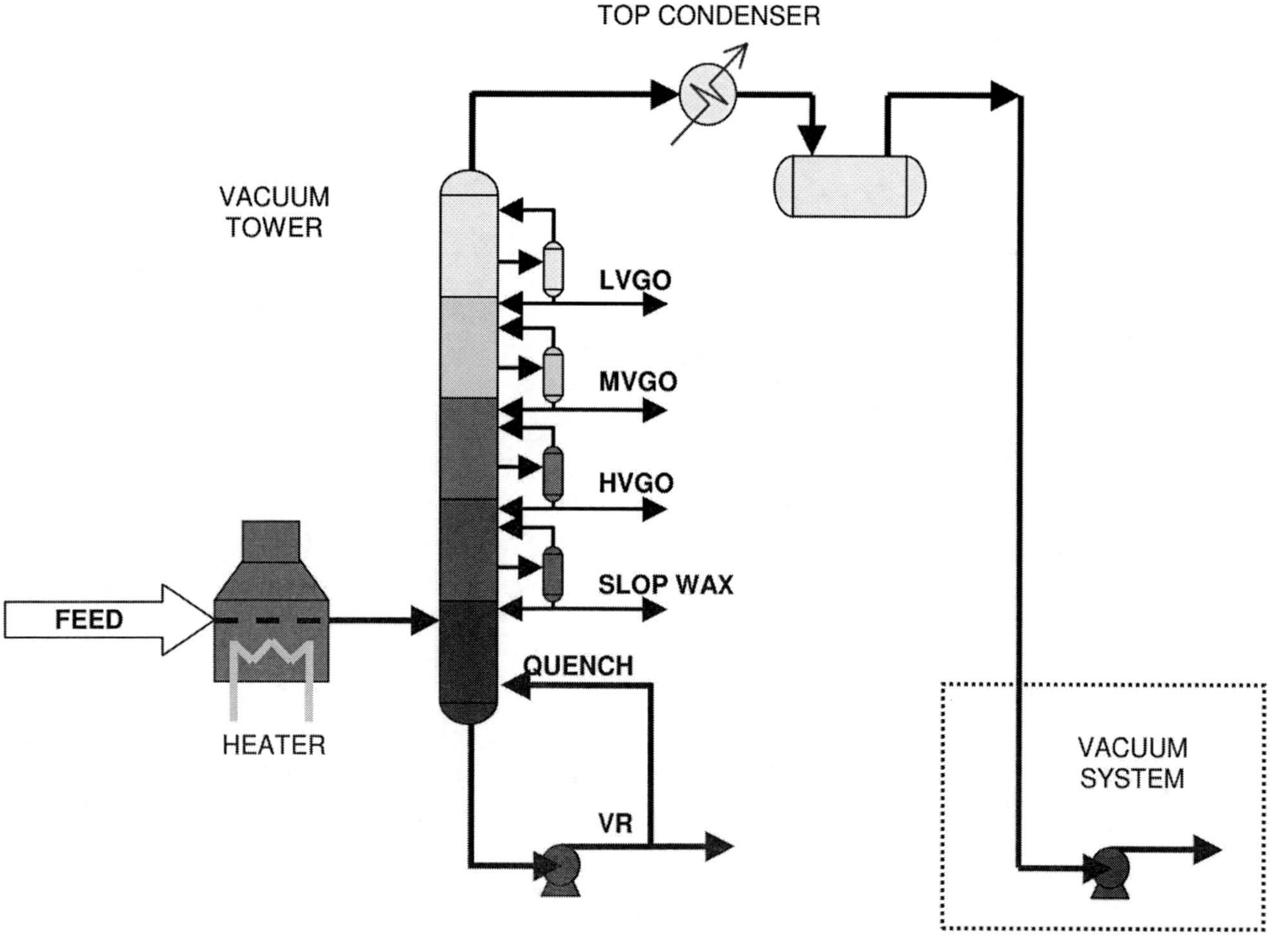

Figure 13 Process Sketch of the Vacuum Pilot Tower

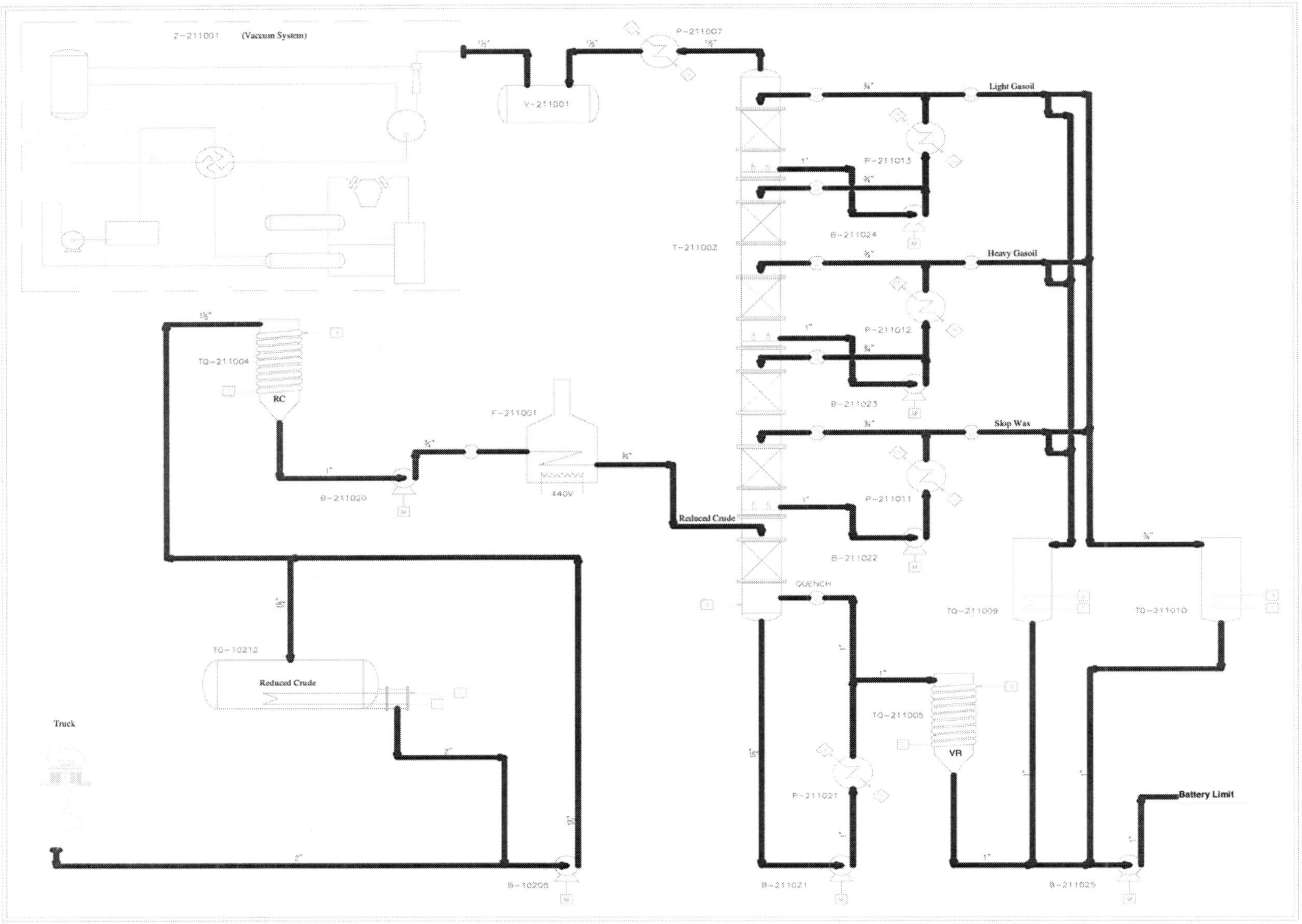

Figure 14 Preliminary PFD of the Vacuum Pilot Unit

Figure 15 Local for Future Installation

Figure 16 Local for Future Installation

Conclusions

To optimize vacuum tower operation it is necessary to cut deeper in order to increase the vaporization and to reduce the vacuum residue yield, and also a good knowledge of the factors that could cause operation failure.

Petrobras decided to build a Vaccum Pilot Plant in order to study different process configurations and internals, as well as metallurgic studies, process simulation validation, and craked gases formation correlations.

One way to increase yield distillates is reducing the tower pressure drop using empty spray sections, achieved by eliminating packings from heat transfer sections. Other benefits are the reduction in the coking tendency and maintenance.

There is a maximum C-factor allowed in empty spray sections. Petrobras developed a fast and simple method to calculate this value from operational data. Based on this methodology two existing towers (REFAP and REMAN) were chosen to have their packing removed.

REFAP vaccum tower revamp will use empty spray sections at the LVGO and dirty wash pumparounds and a new modern vapor horn distributor will be installed.

A small vaccum tower of REMAN refinery was chosen to be Petrobras deep cut benchmark. The base test run was set and new performance tests will be scheduled.

References

Barbabela, B.A.; Guimarães, F.M.Q.G.; Waintraub, S.; Torres, G.A. Optimization of Heat Transfer Zones in Distillation Columns, Presented at X International Phoenics Users Conference, Melbourne, Australia, May 2004.

Cai, T.J.; Kunesh, J.G. Heat Transfer Performance of Large Structured Packing, Presented at the American Institute of Chemical Engineers Spring National Meeting, Houston, TX, March 1999.

Kunesh, J.G. Direct-Contact Heat Transfer from a Liquid Spray into a Condensing Vapor, Ind. Eng. Chem. Res., 32(10), 1993

Ognisty, T.P. The Direct-Contact Heat Transfer Perfomance of a Spray Nozzle, a Notched Trough Distributor and Two Inch Pall Rings, Presented at the American Institute of Chemical Engineers Spring National Meeting, Orlando, FL, March 1990.

Ognisty, T.P.; Kao, Y.S. Performance of Fractionation Devices in Direct Contact Heat Transfer, Presented at Pacific Area Chemical Engineering Congress, Acapulco, Mexico, October 1988.

Waintraub, S.; Torres, G.A ; Guimarães, F.M.Q.G.; Barbabela, B.A. Is it Necessary to Use Internals in Heat Transfer Sections of Refinery Columns ? Presented at the American Institute of Chemical Engineers Annual Meeting, San Francisco, CA, November 2003.

82b Revamping a Sequential Distillation Scheme to a Primary Fractionation Scheme

Daryl Hanson, Mark Myers, Rodney Butts, Steve White

Calumet Lubricants Company operates a solvents plant that produces over 14 separate specialty products. Many of the towers are two-product (top and bottom) fractionators, while one operates with a side-cut product. All of these towers operate in a sequential fractionation scheme where the bottom product from an upstream tower provides the feed to the subsequent tower.

In 2004, Calumet studied the possibility of reconfiguring their solvent train in order to increase capacity and operating efficiency. The results of the study indicated that the existing solvent train could be reconfigured to increase capacity and reduce energy utilization per unit of feed charged. The reconfigured design utilizes a Primary Fractionator to produce five product fractions that are further fractionated in downstream towers. While increasing the capacity and reducing energy consumption, the design will stay within the limitations of the existing column shells. The project schedule for this revamp was very tight. A twenty (20) week schedule from design to installation was accomplished.

82c Investigation of an Improved Strategy for Controlling a Multi-Draw Distillation Column

Eric R Walmet, J George Hayden, Robert S Coeyman

Accurate process control is paramount to the successful operation of a distillation column. In multi-component distillation, such as is present in a tall oil refinery, many traditional techniques for controlling distillations columns are difficult to implement, and simple, yet reliable techniques, such as material balance control typically are utilized. However, in order to improve efficiency in separating components while ensuring that products always meet specification, some advanced control techniques can be implemented.

Steady-state process simulations of our main fractionating column have shown that the mixed acid side stream cut is extremely sensitive to variations in the column and potentially could be used as a control point for maintaining the composition profile within the column. This is largely due to the concentration profile being steeper near the center of the column rather than at the "ends" of the column with the main product draw points.

Early trial work verified that composition of the mixed acid side stream, which is a minor cut, was strongly correlated to composition of both the light acid (upper) side stream (main product draw and current control point) and heavy (lower) side stream cuts. In addition, density of the mixed acid side stream was correlated to composition of both the mixed acid side stream and the light acid side stream. This density was measured on-line via a mass flow meter, which allows for quicker response than obtaining samples and performing the analysis off-line in the lab.

Plant trials have been run with various control strategies to operate the main fractionating column in closed loop control, using the mixed acid density correlation to manipulate the light acid product flow, and thus controlling the material balance in an automated fashion. While a permanent strategy hasn't been implemented, the technique shows promise for the future.

82d Short-cut modelling of Atmospheric and Vacuum Refinery Distillation Columns

Megan Jobson, Robin Smith, Vikas Rastogi

The atmospheric and vacuum distillation columns are the main columns of a refinery crude oil distillation unit. These columns have significantly complex configurations and use live steam for stripping, as water is insoluble with hydrocarbon components. Moreover, the live steam allows separation at lower temperatures compared to reboiling, which prevents cracking of hydrocarbons at high temperatures. These columns have side-strippers or side-rectifiers; pump-around or pump-backs provide local reflux and allow recovery of heat at different temperature levels.

A decomposition approach was developed for simplified modelling of atmospheric distillation columns (Liebmann et al., 1998). The atmospheric column was decomposed into a sequence of thermally-coupled simple columns, including both reboiled and steam-stripped columns. Short-cut methods based on the Fenske-Underwood-Gilliland correlations were developed for grassroots and retrofit design of the simple columns (Suphanit, 1999; Gadalla et al., 2003). These short-cut methods are used along with a simplified model for the existing heat exchanger network to determine energy-efficient, economic and environmentally benign operating conditions for a heat-integrated atmospheric distillation column.

In recent work, a systematic approach is developed to consider the pressure drop in crude oil distillation columns. The validity of the model is illustrated for a typical atmospheric distillation column by comparison of results with rigorous simulation results. The prediction of the temperature profile in the column improves significantly by considering pressure drop; this improves the accuracy with which the heat integration opportunities can be analysed in the heat exchanger network.

A decomposition approach similar to that used for the atmospheric distillation column is proposed for simplified modelling of the vacuum distillation column. Each column in the thermally-coupled sequence is modelled using short-cut methods. To accommodate the features of vacuum distillation columns, two modifications are made:

1. The analytical correlations based on the Gilliland graphical method to calculate the number of theoretical stages for a required separation do not cover the range of vacuum distillation column operation. An analytical expression is developed similar to the Molokanov correlation with modified regression coefficients for vacuum distillation column operation.

2. The method for calculating the distribution of non-key components between the top and bottom products at actual reflux conditions is revised to account for the high reflux ratios in the pump-around sections of the vacuum distillation column.

An illustrative example of a typical vacuum distillation column with two pump-arounds shows that the short-cut simulation results for grassroots and retrofit design compare well with those obtained by rigorous simulation using commercial software. These models for the vacuum distillation column can be used with pinch analysis to explore the potential energy-saving opportunities by exploiting interactions between the distillation columns and the heat exchanger network. The models can be used to determine the operating conditions of the vacuum column to maximise the yield of its distillates. Furthermore, these models can be combined with hydraulic models, for the calculation of column diameters, to assess the effect of increasing throughput on the hydraulic capacity of an existing vacuum distillation column.

References:

Liebmann, K., V. R. Dhole, and M. Jobson, Integrated design of a conventional crude oil distillation tower using pinch analysis, Chem. Eng. Res. Des., 76(A3), 335-347 (1998)

Suphanit, B., Design of Complex Distillation System, PhD Thesis, UMIST, Manchester, UK (1999)

Gadalla, M., M. Jobson, and R. Smith, Shortcut models for retrofit design of distillation columns, Chem. Eng. Res. Des., 81(A8), 971-986 (2003)

82e Atmospheric and Vacuum Crude Tower Revamp Case Studies

Barbara A Todd, Luz N Hernandez

Refinery case studies on recent crude oil atmospheric and vacuum tower revamps.

82f De-Bottlenecking a HF Alkylation Unit by Retrofitting a De-Isobutanizer with VGPlus Trays

Giuseppe Mosca

Increasing throughput in major existing equipment, such as distillation columns, is a key factor to economically meet production targets. To maximize plant's performances it is mandatory to fully understand the equipment being used, their mechanism of operation, limitations and capabilities. This includes, among others, furnaces, exchangers, pumps, and mass transfer components for distillation towers. The purpose of this paper is to provide an overview of the key factors to succesfully implement revamping projects with the goal of maximizing capacity without compromizing efficiency. A commercial experience will be discussed in details, where the use of one of the most advanced fractionation device, i.e. VGPlus tray, yields to the best cost effective solution of revamping a De-Isobutanizer at a major European Refinery. The paper will cover the description of the unit, the scope of the revamping, its implementation, and the achieved benefits.

Simulation and Analysis of Energy-saving Technology for Absorption-stabilization System

ZHOU Wenjuan[1], *JIANG Bin*[1, 2] *and LI Xingang*[1, 2]
1 Shool of Chemical Engineering and Technology, Tianjin University, Tianjin 300072, China
2 National Engineering Research Center for Distillation Technology, Tianjin 300072,China

Abstract: A new energy-saving technology named second condensation is proposed in this paper. The technology has an especial characteristic that is, after being mixed, compressed rich gas, absorption oil and desorption gas are cooled and then enter the surge tank. The condensed oil is not cooled and is sent to the upper part of the desorption column as heat feeding. The non-condensable gas is cooled secondly, then the condensed oil from the second surge tank is sent to the top of the desorption column as cold feeding. This technology depresses the cold loads before the surge tanks and owns the characteristic of double feeding process. An other new technology that is the combination of intermediate reboiler and second condensation is also proposed in this paper. It adequately embodies many virtues of these two kinds of technologies. Compared with the general technology of two-tower process, the new technologies have prominent superiorities.

Keywords: absorption-stabilization system, FCC, energy-saving, second condensation

1.PREFACE

Fluid catalytic cracking (FCC) process is an important process in secondary processing in oil refining industry in which it plays a very important part. Generally the FCC complex includes the following three sections: reactor and regenerator system, main fractionator system and absorption-stabilization system[1-2].

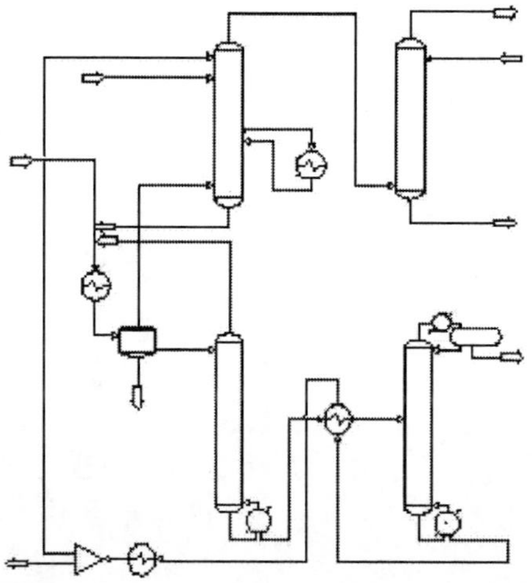

Figure 1 Absorption-stabilization system process chart

The absorption-stabilization system is used to process naphtha and rich gas from the top gas-oil separator of fractionator, separate dry gas (C_2 and C_2^-) and reclaim LPG(C_3,C_4) and stable gasoline (C_4^+).The absorption-stabilization units mostly consist of absorption column, resorption column, desorption column and stabilizer column[1-3]. Since the 1980s, many refineries have changed single-tower process into two-tower in order to avoid the high content of C_3 in lean gas and C_2 in deethanize gasoline. The operational way of desorption column has very important influence upon the system[3]. The primary feeding ways include cold feeding, heat feeding and double feeding[3-4]. The topical process (cold feeding process) is shown in Fig.1. Tianjin university proposed intermediate reboiler technology which adequately embodies the virtues of cold feeding and heat feeding.

The petroleum industry paid an attention to the problem on energy-saving following the development of economy. A new energy-saving technology named second condensation is proposed in this paper. The technology will be simulated and analysed by adopting PRO/ II software and will be compared with the general technologies of two-tower process.

2.SUMMARY OF TWO-TOWER PROCESS

2.1 Cold feeding process

Cold feeding process is shown in Fig.2. Its characteristic is that condensed oil is sent to the top of desorption column without being heated. The virtue is that it brings the low content of C_3^+ in lean gas, the good effect of absorption, and the little flux of desorption gas, but the defect is that the loads of desorption column and the reboiler of its bottom are large[3].

2.2 Heat feeding process

Heat feeding process is shown in Fig.3. Its characteristic is that condensed oil is sent to the top of desorption column after being heated. It can decrease the loads of reboiler, but the effect of absorption is bad, the flux of desorption gas is increscent and the loads of cooler and condenser also increase[3].

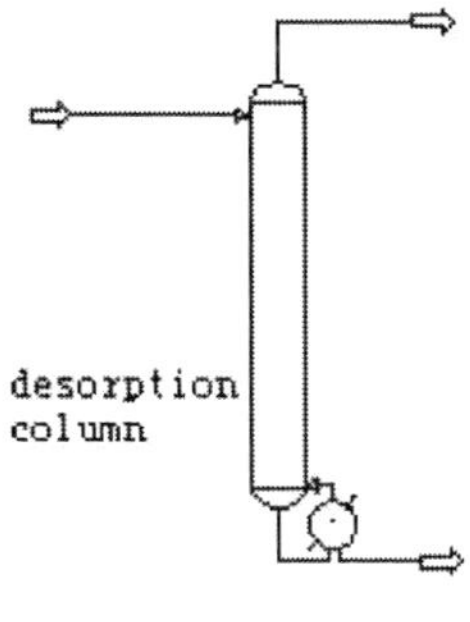

Figure 2 Cold feeding process

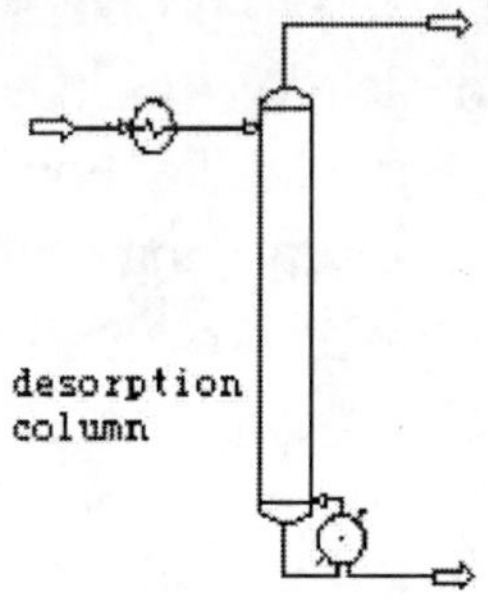

Figure 3 Heat feeding process

2.3 Double feeding process

Double feeding process is shown in Fig.4. Condensed oil from the surge tank has been divided into two sections. One is sent into the upper part of desorption column after heat exchanging with stable gasoline as heat feeding. The other directly enters the top of desorption column as cold feeding. Double feeding process synthesizes the virtues of cold feeding process and heat feeding process and has the obvious superiority. It reduces not only the flux of desorption gas relatively, but also the loads of the reboiler[3].

2.4 Intermediate reboiler process

Intermediate reboiler process is shown in Fig.5. Cold condensed oil entirely enters the top of desorption column, where is a reboiler in the intermediate section. The technology adequately synthesizes the virtues of cold feeding process and heat feeding process and has the obvious superiority. And it also avoids back mixing consisting in double feeding process[4].

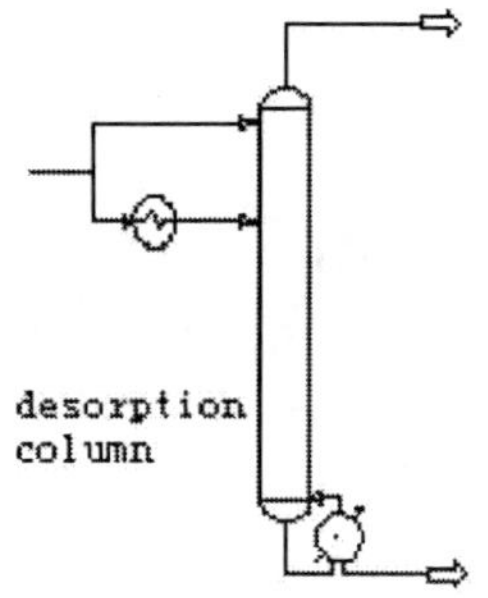

Figure 4 Double feeding process

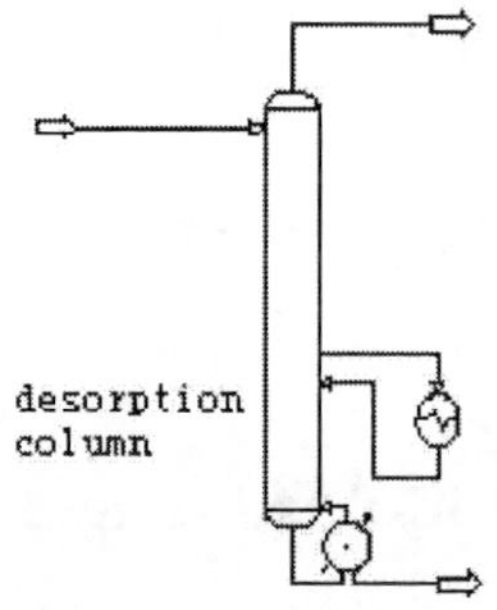

Figure 5 Intermediate reboiler process

2.5 Second condensation process

Second condensation process is shown in Fig.6. The technology has its own characteristic. After being mixed, compressed rich gas, absorption oil and desorption gas are cooled and then enter the surge tank. The condensed oil is not cooled and sent to the upper part of the desorption column as heat feeding. The non-condensable gas is cooled secondly, then the condensed oil from the second surge tank is sent to the top of the desorption column as cold feeding. The technology only makes partially feeding cold, so the cold loads before the surge tank can be reduced to save energy. It possesses the characteristic of double feeding process. But compared with double feeding process, heating the feeding of desorption column is needless.

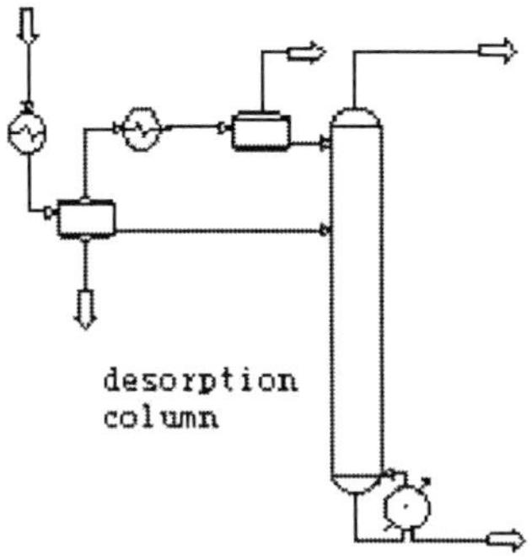

Figure 6 Second condensation process

2.6 The combinative process of second condensation and intermediate reboiler

The combinative process of second condensation and intermediate reboiler is shown in Fig.7. The combinative process has double characteristics. Second condensation reduces the cold loads before the surge tank, and intermediate reboiler reduces the heat loads of reboiler at the bottom of desorption column. This process adequately combines the virtues of cold feeding and heat feeding.

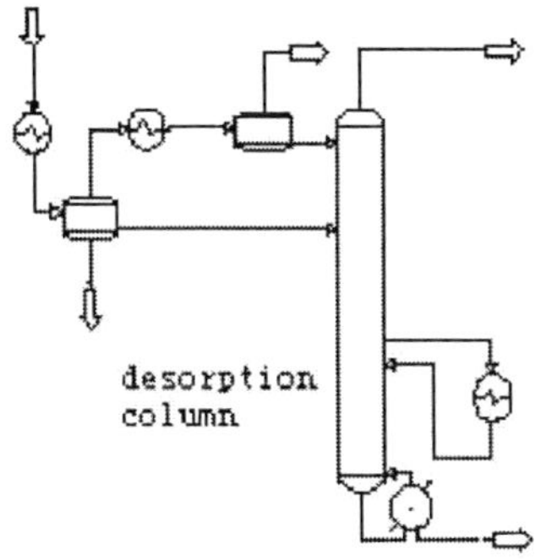

Figure 7 The combinative process of second condensation and intermediate reboiler

3.BASIC DATA AND QUALITY CONTROL INDEXES

(1) The catalytic working ability of certain refinery is 80Mt/y. The flux of compressed rich

gas is 14500kg/h, naphtha is 28000kg/h, and light diesel oil is 6000kg/h.

(2) Tower top pressure of absorption tower is 1.1MPa (absolute pressure), differential pressure is 0.03MPa and operating temperature is 313K. Tower top pressure of desorption tower is 1.24MPa and differential pressure is 0.04MPa.

(3) N.T.P of absorption column is 15. The column has two-stage intermediate heat exchange. First stage cooler is between seventh tray and eighth tray. And second stage cooler is between eleventh tray and twelfth tray.

(4) N.T.P of absorption column is 20. The temperature of cold feeding is 40℃. The temperature of heat feeding is 75℃. As double feeding, the temperature of cold feeding is 40℃ and the temperature of heat feeding is 78℃. Intermediate reboiler has an intermediate heat exchange between fourteenth tray and fifteenth tray, and the return temperature is 100℃. When it is second condensation, the temperature of first condensation is 65℃ and that of second condensation is 40℃. When second condensation and intermediate reboiler combines, the temperature of first condensation is 58℃.

(5) Volume composition of rich gas is shown in Table.1.

Table 1 Volume composition of the rich gas

order number	Component	V%	order number	Component	V%
1	H_2	16.23	11	C_3H_6	9.14
2	N_2	6.22	12	nC_4	2.76
3	CO	0.44	13	iC_4	8.37
4	CO_2	1.37	14	$1\text{-}C_4=$	2.94
5	O_2	0.89	15	$i\ C_4=$	3.91
6	H_2S	0.25	16	$CC_4=$	2.66
7	CH_4	10.65	17	$TC_4=$	3.18
8	C_2H_6	7.13	18	nC_5	3.17
9	C_2H_4	7.35	19	iC_5	4.95
10	C_3H_8	5.14	20	$1\text{-}\ C_5$	5.25

(6) Data of D86 distillation curve of crude gasoline and light diesel oil are shown in Table.2.

Table 2 Data of D86 distillation curve of crude gasoline and light diesel oil

	crude gasoline	light diesel oil
HK	/	148
5%	20	207
10%	50	226
30%	83	256
50%	105	272
70%	141	294
90%	178	325
95%	188	335
KK	193	348
ρ / kg/m3	709	806

(7) Quality control indexes:

Dethanizing gasoline: C_2<0.1% (mol)

LPG: C_5<0.1%(mol)

Stable gasoline: C_4<0.5%(wt)

4.COMPARISON OF SIMULATION RESULTS

In order to compare effectively, one benchmark is chosen. The benchmark is the composition of C_3^+ in dry gas is 1.5%(mol).

Comparison of the cold loads is shown in Table.3. To the cold loads before the surge tank, if the loads of cold feeding process is a benchmark, the loads of heat feeding increases by 12.5%, while double feeding and intermediate reboiler increase by 1.8% and 1.5%, respectively. To second condensation process, compressed rich gas, absorption oil and desorption gas are firstly cooled to 65℃ and are sent to the surge tank after being mixed. The condensed oil which is in the majority is regarded as heat feeding. A small quantity of non-condensable gas is secondly cooled to 40℃ as cold feeding. So compared with cold feeding process, the cold loads of second condensation process decrease by 57.8%. From Table.3, we know that the technology needs more loads of intermediate heat exchange, because there is high content of $C3^+$ in desorption gas. In order to make the percentage composition of $C3^+$ in dry gas at 1.5%, the quantity of heat exchange must be increased. The cold loads before the surge tank of the combinative process decrease by 40.5%. For the total cold loads, cold feeding process has maximum, and the other processes decrease by upwards of 8%, especially the combinative process that decreases by 13.6%. It is obvious that second condensation process and the combinative process have prodigious superiority when considering the energy-saving.

Table 3 Comparison of the cold loads (M*kJ/h)

process	Cold feeding	Heat feeding	Double feeding	Intermediate reboiler	Second condensation	The combinative process
Cold loads before surge tank	-6.911	-7.778	-7.038	-7.018	$-2.919\left\{\begin{array}{l}-1.652(first)\\-1.267(second)\end{array}\right.$	$-4.085\left\{\begin{array}{l}-3.416(first)\\-0.669(second)\end{array}\right.$
Loads of first stage cooler	-0.328	-0.450	-0.300	-0.268	-0.986	-0.850
Loads of second stage cooler	-0.150	-0.270	-0.120	-0.126	-0.631	-0.385
Colds load of stable gasoline	-8.342	-3.980	-5.932	-5.037	-8.858	-6.490
Loads of condenser of stabilizer column	-12.510	-12.760	-12.590	-12.589	-12.575	-12.583

Comparison of the heat loads is shown in Table.4. Because the heat quantity of the latter three items is from stable gasoline in the system, the former two items are only considered. For the loads of the reboiler of desorption column heat feeding process which is compared with cold feeding process decreases by 30.7%. The latter four processes decrease by 23.4%, 32.1%, 29.9% and 38.2%, respectively. Comparing the total heat loads, cold feeding has maximum, the other processes decrease by 13.7%, 10.4%, 14.6%, 13.5% and 17.6%, respectively. This shows that the combinative process is better than the others in this aspect. Double feeding process is the worst except cold feeding process.

The character of desorption gas is shown in Fig.8. If cold feeding process is a benchmark, the flux of desorption gas of heat feeding process increases by 102.1%. There is a great deal of C_3 and C_4 composition in desorption gas which increases by 193.3% and 224.5% respectively. The flux of C_3 in desorption gas of double feeding process increases by 14.6%. The result of intermediate reboiler process is as much as cold feeding process. The flux of desorption gas of second condensation process reduces by 4.9%, but the flux of C_3 and C_4 composition increases by 37.6% and 58.9%. This results from that 95.4% feeding is heat feeding and more C_3 and C_4 compositions circulate in desorption gas. The flux of C_3 and C_4 composition of the combinative process is less than second condensation process. Thus heat feeding process is incomparable; intermediate reboiler process has better effects on the character of desorption gas; there is the least desorption gas in second condensation process, but the flux of C_3 and C_4 composition is biggish. The combinative process of second condensation and intermediate reboiler is intervenient.

Table 4 Comparison of the heat loads (M*kJ/h)

process	Cold feeding	Heat feeding	Double feeding	Intermediate reboiler	Second condensation	The combinative process
Loads of reboiler of desorption column	10.611	7.352	8.128	7.204	7.437	6.553
Loads of reboiler of stabilizer column	11.458	11.697	11.643	11.643	11.649	11.644
Heating loads of heat feeding	—	4.730	2.908	—	—	—
Loads of intermediate reboiler	—	—	—	3.783	—	2.346
Preheating loads of dethanizing gasoline	6.273	6.251	6.125	6.132	6.089	6.122

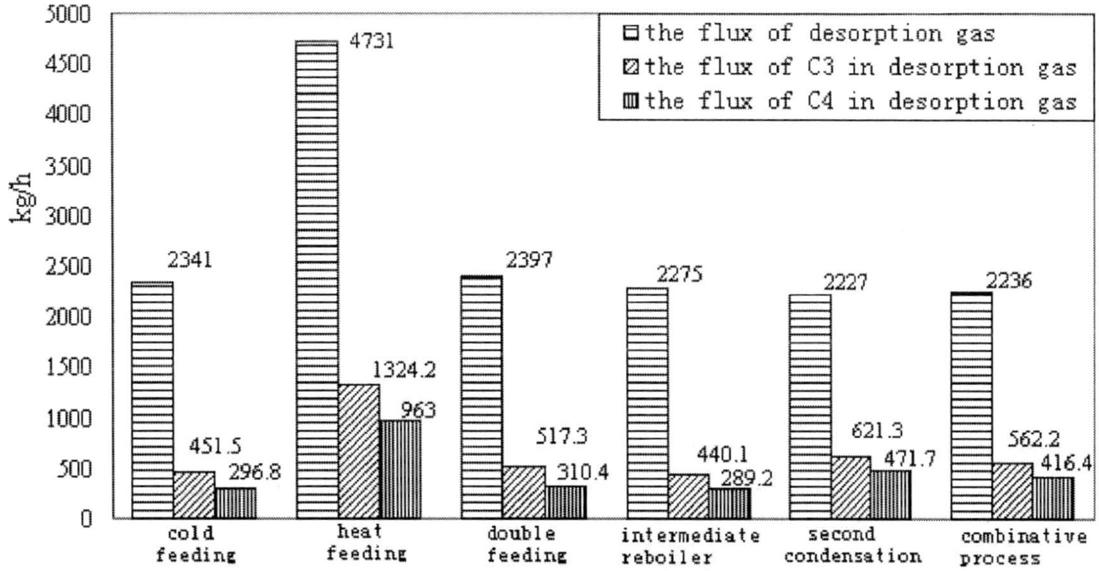

Figure 8 Flux and content of the desorption gas (kg/h)

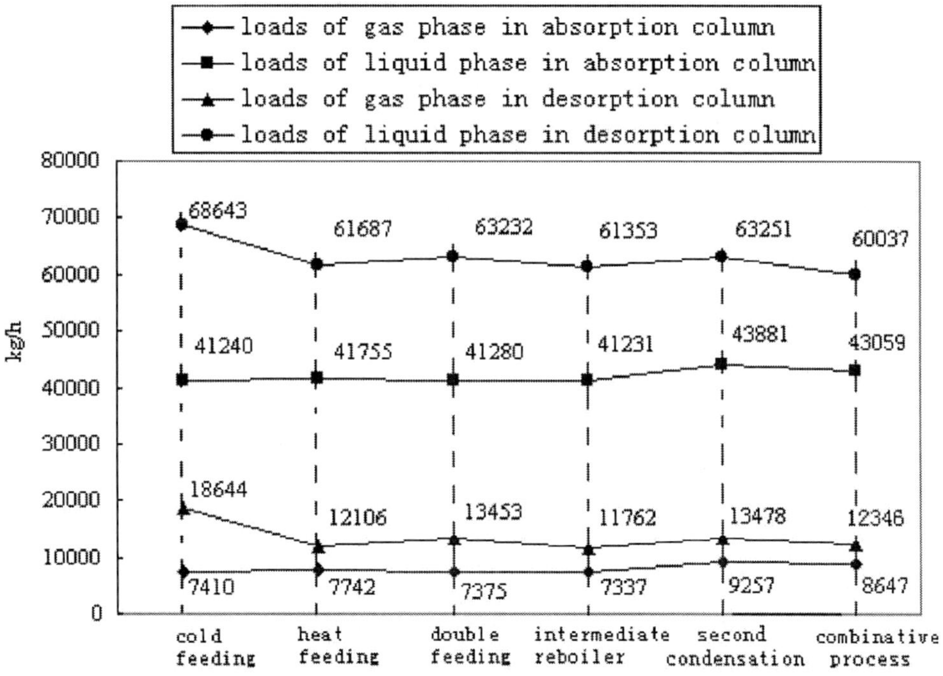

Figure 9 Maximum load in the columns（kg/h）

Maximum load in the columns is shown in Fig.9. Cold feeding process is regarded as a benchmark. The load of gas phase in absorption column: second condensation process increases by 24.9%; the combinative process increases by 16.7%; heat feeding process only increases by 4.5%; double feeding and intermediate reboiler are basically comparative with cold feeding process. The load of liquid phase in absorption column: the former four processes are comparative; compared with cold feeding process, the latter two processes only increase by 6.4% and 4.4%. The load of gas phase in desorption column: that of cold feeding is maximal; the loads of the other processes all decrease, especial the combinative process that decreases by 33.8%. The load of liquid phase in desorption column: the load of cold feeding is maximal; the loads of the other processes all decrease, especially the combinative process that decreases by 12.5%. So we can obtain the following results: intermediate reboiler process is better; second condensation process makes the load of gas phase in absorption column large, but there are good effects on the other aspects; maximum load in absorption column of the combinative process is less than second condensation process, and maximum load in absorption column is the least.

5.CONCLUSIONS

This paper has simulated some technologies of two-tower process for absorption-stabilization system. In order to compare, a benchmark is drafted in the course of simulation. That is the percentage composition of C_3^+ in dry gas is 1.5%(mol). And the percentage composition of C_3^+ is adjusted by changing the quality of heat exchange in absorption column. The conclusions are as follows:

(1) A new energy-saving technology named second condensation is proposed in this paper. The cold loads before the surge tank of this technology is very little, so it can save energy. In this example, that of second condensation process and the combinative process decreases by 57.8% and 40.5%.

(2) In the six technologies, from the charter of desorption gas heat feeding process is the worst and intermediate reboiler process is the best. Compared with double feeding process, second condensation process has comparative flux of C_3 in desorption gas; compared with cold feeding process, it has more flux of C_3 and C_4, but the flux of desorption gas is the least. The effect of the combinative process is between second condensation and intermediate reboiler.

(3) From cold and heat loads second condensation and the combinative process are beyond comparison. Not only cold loads but also heat loads have prodigious superiority on energy-saving. But the effect of cold feeding and heat feeding process is worse.

(4) Compared synthetically from heat and cold loads, the character of desorption gas and maximum load in the columns, these technologies are described as follows: heat feeding process is the worst of all; intermediate reboiler process is the best on the character of desorption gas and maximum load in the columns; second condensation process and the combinative process have absolute advantages on saving cold loads and heat loads.

REFERENCES

1 Robert A Meyers, Handbook of petroleum refining processes, 2-10-2-22 (1986)

2 Lu Qingyun, Fluid Catalytic Cracking (in Chinese), 263-288 (1989)

3 Li Xingang, Sun Jinsheng, Jiang Bin and Zhao Ruwen. Flowsheet simulation and for FCC absorption/stabilization unit. Petrochemical Engineering Design (in Chinese), 13(2):45-53 (1996)

4 Du Xiang, Wu Shaomin, Li Changgeng, Du Yingsheng et al. Chemical Engineering(in Chinese), 33(3):16-21 (2002)

82h CFD-based Multi-scale Method for Predicting Mass Transfer Performance in Structured Packing Columns

Chunjiang Liu, GU Fang, K T Yu, Mike D Pritchett, Xigang Yuan

It was well known that the separation efficiency of distillation process performed in structured packing columns depends strongly on the quality of initial distributions of liquid and vapor phase. It means that the fluid dynamics of the phases was a dominant factor in determining the column performance and quantitative investigation of its effect becomes critical for predicting the mass transfer efficiency of the column. Up still now, most of the published work on predicting the column separation performance was assumed plug flow for both phases and methods proposed in these works are not conservative enough to insure safe designs. In this paper we try to propose a CFD-based multi-scale method to predict the mass transfer performance for structured packing columns. At present, some single-phase flow or some simple two-phase flow problems can be simulated perfectly by CFD techniques. But direct and detailed CFD simulation of the fluid dynamics for the whole column packed with structured packing was impossible under the currently available computing power. Therefore, CFD-based multi-scale method which include a crunodes meshwork model for predicting the macro-scale liquid distribution in structured packing columns, a CFD two-phase rivulet film flow model for describing the local rivulet film flow patterns on the structured packing surface, and a gas phase CFD model for calculating micro-scale gas distribution is established. Since the locally two-phase flow patterns can be obtained directly by the local CFD models, a lot of empiristic correlation formulae used in traditional methods, such as effective area, mass transfer coefficient, local gas or liquid velocity etc., can be abandoned in present method and only some essential liquid or gas properties and packing geometries was needed. The simulated results by the proposed method show a good agreement with the experimental data under the condition of normal or vacuum pressure.

Keywords: structured packing column; mass transfer performance; computational fluid dynamics(CFD); multi-scale model,

83a Nonlinear Model Predictive Control of a Reactive Distillation Column

James Riggs, Rohit Kawathekar

Reactive distillation combines both separation and reaction in one unit. It has been used in a small number of industrial applications for many years. Reactive distillation can offer significant economic advantages in some systems, particularly for reversible reactions which are limited by equilibrium. Reactive distillation is a challenge for control due to process nonlinearity, complex interactions between vapor-liquid equilibrium, vapor-liquid mass transfer and chemical kinetics. This paper considers the application of nonlinear model predictive control (NLMPC) and conventional PI controls to a highly nonlinear reactive distillation column in a two-column process for producing a high purity ethyl acetate product. The steady-state process gains are shown to change between 33 and 600% for approximately a 10% change in the manipulated variables. This process was designed to optimally produce high purity ethyl acetate from a steady-state perspective, but results in a challenging control problem.

A detailed tray-to-tray distillation model with reaction considered on each tray was used to represent the ethyl acetate reactive distillation column considered in this study. NLMPC was applied as a nonlinear programming problem using orthogonal collocation on finite elements to approximate the ODE's that constitute the model equations for the reactive distillation column. Diagonal PI controls were used to identify that the [L/D,V] and the [L/D/V/B] control configurations performed best. The effect of the various NLMPC tuning parameters and design decisions were investigated to ensure that optimal performance for the NLMPC controller was attained. The PI controllers were tuned using Auto Tune Variation along with optimal tuning using setpoint changes.

NLMPC was applied using the [L/D,V] configuration and found to provide a factor of 2-3 better performance than the corresponding PI controller. The effect of process/model mismatch on the performance of the NLMPC controller was also evaluated. Dynamic decoupling between the overhead and the bottoms product resulted in reasonable performance by the PI controllers.

83b A Novel Concept: Semicontinuous Reactive Distillation

Thomas A Adams, Warren D Seider

We introduce a novel method of combining reaction and distillation unit operations by using semicontinuous distillation methods with a CSTR as one of the middle vessels. While normal reactive distillation uses catalysts inside a column to perform a reaction and a separation concurrently, semicontinuous reactive distillation (SRD) performs a similar task, and is a plausible alternative design scheme for certain types of process goals. The semicontinuous method is best chosen for systems that need production rates higher than a batch system would support economically, but low enough that using a continuous system would be cost prohibitive. The SRD design scheme should also be considered for reaction systems where the kinetics are too slow to operate a reactive distillation column efficiently, particularly where reactions are highly reversible, or where product boiling points are too close together to separate efficiently with regular distillation.

The SRD system studied in this paper considers the reaction of Acetaldehyde and Propylene reacting reversibly to form 2,4-dimethyl-1,3-dioxolane (DMD) and water. This reaction system is interesting because the products form a low boiling azeotrope, where the products and the azeotrope have boiling points in between that of the two reactants. Moreover, DMD and water have boiling points that are very close together. Because of the narrow boiling point range, the azeotrope, the inability to control the reaction temperatures, and the significant reversible reaction, ordinary reactive distillation is very difficult. Thus, the SRD system becomes a viable candidate.

The SRD system operates in a cycle of different modes, where products and remaining reactants are withdrawn from the CSTR and separated, recycling one of the reactants back to the reactor and accumulating the balance in a tank. Then, new feed is charged to the CSTR, while the column switches to a new mode where it is used to purify the products in the accumulation tank. Because of the great shift in column profiles, it is necessary for the system to have several short transition modes where the column operates at total reboil while undesired chemicals are removed from the column and the compositions shift to new desired values.

The results of the study show that it is possible to use SRD to obtain arbitrarily pure products from a difficult reaction system using a relatively small amount of process equipment. It is shown that given a regular supply of feed, the SRD system converges to a consistent cycle where purified product can be obtained at regular intervals with minimal reactant losses. Thus, by reusing process equipment for multiple purposes and performing tasks in parallel, the SRD system can work around problems that traditional processes cannot overcome efficiently.

83c Pressure Sensitivity in Distillation - Implications for Control Structure

Sten B Jorgensen

Modelling of distillation columns often neglects secondary effects such as pressure sensitivity of the separation. In this paper it is demonstrated that such effects may be important for the behaviour of even fairly simple columns if they are not properly accounted for. A specific effect related to the pressure sensitivity of the separation is investigated both through simulation and experiment. It is demonstrated that it can be important where in the column pressure is controlled even for binary separations. The finding also has relevance for floating pressure columns depending on the specific pressure sensitivity of the separation.

Through insight and simulation it has been suggested that there exists an input multiplicity in binary distillation columns for a specific but common type of separation. The work reported here aims at demonstrating this input multiplicity experimentally in a distillation pilot plant and at revealing the implications thereof for pressure control structure selection in binary distillation. Opposing effects of the internal flow rates on separation cause the input multiplicity. I.e. an increase in internal flow rates will change the slope of the operating lines with a positive impact on separation. However, increased flow rates will cause an increase in the column pressure, which will change the vapour-liquid equilibrium with –most often - a negative impact on separation. For relatively low flow rates the improved operating lines will dominate the effect on vapour-liquid equilibrium. However, at relatively high flow rates the effect on vapour-liquid equilibrium will dominate. I.e. the gain between internal flows and separation will change sign and cause an input multiplicity. The influence of entrainment is also discussed.

Quite elaborate experimental work on a heat integrated distillation pilot plant has been used to verify the input multiplicity. The basic idea is to obtain steady-state profiles for different values of the internal flow rates while maintaining other operational parameters constant. Experiments are carried out for different feed rates and also using two different points for controlling column pressure. The observed results strongly suggest the presence of input multiplicity for the relatively common specific type of separation property.

The consequential control point for column pressure depends upon the separation property. For most mixtures pressure should be controlled in the reboiler, to avoid the unpleasant effects of the input multiplicity, when using a decentralized control structure for the column. The effect of controlling pressure in the column top versus bottom is also investigated experimentally.

83d Evaluation of the Separation Efficiency of Three-Phase Operated Packed Towers

Günter Wozny, Jens-Uwe Repke, Olivier Villain

In past and recent studies ([1], [2]), the understanding of the mechanism governing the three-phase distillation with packing was significantly improved. Nevertheless, the predictive calculation of the separation efficiency of three-phase operated packing is still not supported by the simulation, and existing models are generally not convenient for common industrial strategic tasks. Therefore, new modelling methods for the consideration of a second liquid phase on the packing surface are needed, and also new designing methods are needed to improve the industrial application of packed towers for the three phase distillation. In order to achieve short-term results, a new approach to the investigation of three-phase distillation with packing was developed and presented in [3]. The information resulting from this study not only enables the determination of decisive parameters for the prediction of the separation efficiency, but constitutes also a solid base for the validation of future theoretical works. In this contribution the results of the investigation of the efficiency of three-phase operated packing based on systematic experiments and a case study are presented.

Experimental results

Because of the abundance of parameters influencing the separation efficiency, the goal of this study is to simplify the problem by identifying three phase distillation classes. Thus, the main focus of the experimental approach is to find out determining dependencies for the distillation performances: for example the type of packing or some physical properties of the considered mixture (density, viscosity, scale of the miscibility gap, surface tension). To achieve this target, a comprehensive experimental study was carried out and representative data for most of the possible three-phase distillation cases were produced. About 500 data sets were obtained during the study. In the presentation, the complete results of the experimental investigation will be presented. Furthermore, a detailed method to determine the relevant parameters for forecasting the separation efficiency of three-phase operated packing will be proposed.

Model considerations

A model is developed for the prediction of the separation efficiency of a three-phase distillation process in packed towers. The model is based on the calculation of the Height of a Transfer Unit (HTU) of a three-phase operated packing for the multi-component mixture case. The method proposed in [4] is considered for the calculation of the multi-component mass transfer coefficients, whereas new methods are proposed to scale those coefficients for one component and to take into account the occurrence of a second liquid phase on the packing. The binary mass transfer coefficient calculation is taken from [5]. Regarding the different aspects of the modelling, the paper will focus on the validation of the conception of the model. Comparison between simulation and experiments will be presented.

Conclusion

The comprehension of the relationship between separation efficiency, design and operating conditions for various fluid systems and for various packing represents a fundamental step in the investigation of heterogeneous distillation. Therefore, the results of the systematic experimental investigation of three-phase operated packing are considered for the development of short-cut evaluation methods. In addition, the paper focuses on the validation of a short-cut method to support the evaluation of the three-phase distillation performance. This short-cut method is based on a predictive model to calculate the Height of a Transfer Unit for the multi-component three-phase distillation. Thus, the developed evaluation technique allows prediction of the separation efficiency for wide range of distillation cases.

References

[1] Harrison, M .E.: *Consider Three-Phase Distillation in Packed Columns*, Chem. Eng. Prog. 86 (1990) 11, pp. 80 - 85

[2] Repke, J.-U.; Wozny, G.: *Experimental investigations of three-phase distillation in a packed column*, Chemical Engineering Technology. 25 (2002) 5, pp. 513 -519

[3] Villain, O.; Repke,J.-U.; Wozny, G.: *Performance characterization of three-phase operated packing*, AIChE annual 2003, San Francisco, 16-21 November

[4] Taylor, R.; Krishna, R.: *Multicomponent Mass Transfer*, John Wiley & Sons, inc., New York, 1993

[5] Billet, R.: *Packed Towers*, VCH Verlagsgesellschaft GmbH, Weinheim, 1995

Crossing Distillation Boundaries in Synthesis of Separation Sequences for Azeotropic Mixtures

Stanislaw K. Wasylkiewicz

Aspen Technology, Inc.
900, 125 - 9th Avenue SE, Calgary, Alberta T2G 0P6, Canada

Presented in the Distillation Symposium of the 2005 Spring AIChE Meeting,
Atlanta, GA, April 10-14, 2005, Paper No. 83e

Outline

- Introduction
- Structure of Separation Space
- Distillation Regions and Boundaries
- Generation of Distillation Regions
- Primary and Secondary Recycles
- Synthesis of Separation Sequences
- Conclusions

Introduction

- This work is part of our development in **DISTIL**, an analysis, design and synthesis toolkit for distillation systems.
- The objective of this work was to develop an integrated and systematic approach to distillation synthesis, which will allow the process engineer to identify the best design alternatives, improve process understanding and reduce overall design time.

Introduction

Thermodynamic tools:

- rigorous thermodynamic models (EOS and activity coefficient models)
- comprehensive thermodynamic databases (e.g. TRC)
- parameter estimation and regression from experimental data
- rigorous phase equilibrium and stability calculations
- finding all azeotropes, topological consistency check for the system, pressure analysis of azeotropes
- automatic creation of RCMs and VLLE diagrams

Introduction

Synthesis tools:

- identification of distillation regions and distillation boundaries
- rigorous feasibility test and split generation method
- crossing curved distillation boundaries
- non-isobaric distillation column sequences
- decanter splits
- identification of suitable recycling options
- recycle mass balances

Framework for Synthesis of Separation Systems

- Prepare a rigorous thermodynamic model of the mixture, estimate and regress parameters from experimental data[1]
- Calculate all azeotropes[2,3] and adjacency and reachability matrices[4]
- Perform pressure analysis of azeotropes in appropriate pressure range
- Find all basic distillation regions[4] and then divide them into continuous and batch distillation regions[5]
- Determine feasible products from individual separators
- Identify recycle options and calculate mass balances
- Do preliminary design and estimate costs

1 Wasylkiewicz, S. K., L. C. Kobylka, and M. A. Satyro, Chemical Engineering, 80-85, August 1999.
2 Fidkowski, Z. T.; Malone, M. F.; Doherty, M. F. Comput. Chem. Eng. 1993, 17, 1141-1155.
3 Wasylkiewicz, S. K., M. F. Doherty, and M. F. Malone, Ind. Eng. Chem. Res., 38, 4901-4912 (1999).
4 Rooks, R. E., V. Julka, M. F. Doherty, and M. F. Malone; AIChE J. (1998).
5 Wasylkiewicz, S. K. and F. J. L. Castillo, 54 CChEC, Calgary 2004, paper No 497.

Structure of Separation Space

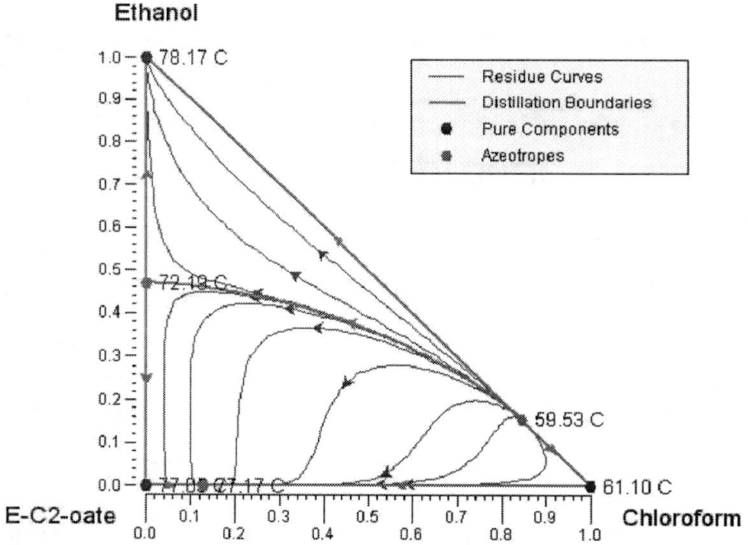

Singular Points (101.325 kPa):

1: Chloroform - Ethanol (59.53 °C)
2: Chloroform (61.10 °C)
3: Ethanol - E-C2-oate (72.19 °C)
4: E-C2-oate (77.05 °C)
5: Chloroform - E-C2-oate (77.17 °C)
6: Ethanol (78.17 °C)

Adjacency matrix:

	1	**2**	**3**	**4**	**5**	**6**
1	0	1	1	0	0	1
2	0	0	0	0	1	0
3	0	0	0	1	0	1
4	0	0	0	0	1	0
5	0	0	0	0	0	0
6	0	0	0	0	0	0

Reachability matrix:

	1	**2**	**3**	**4**	**5**	**6**
1	1	1	1	1	1	1
2	0	1	0	0	1	0
3	0	0	1	1	1	1
4	0	0	0	1	1	0
5	0	0	0	0	1	0
6	0	0	0	0	0	1

Basic Region

Basic Region:
a subspace of a composition space that contains only one unstable node (UN), one stable node (SN) and any number of saddles (SA)

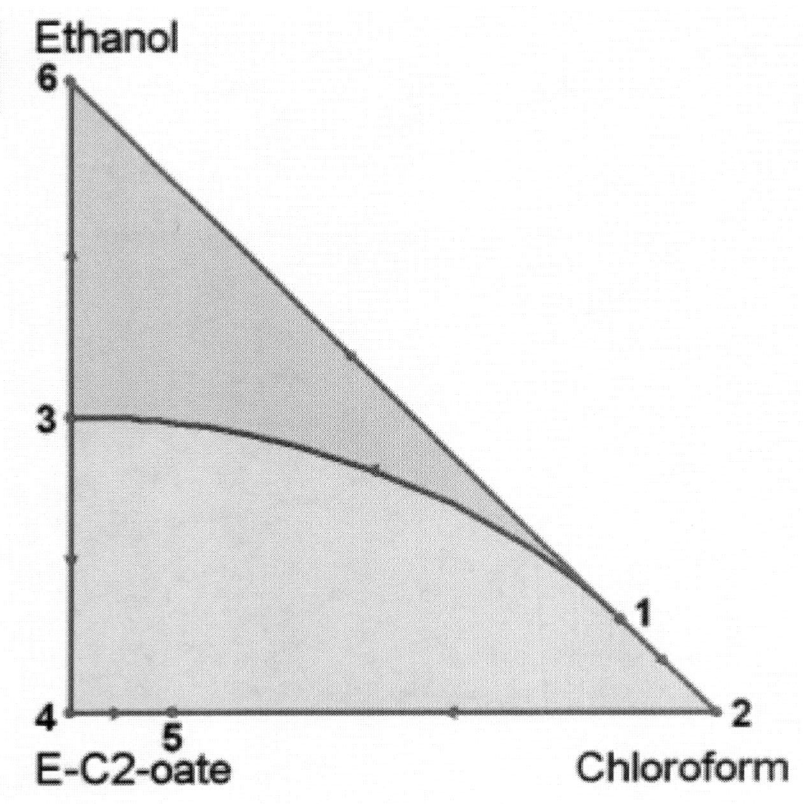

Basic Distillation Regions:

1: 1 2 3 4 5

2: 1 3 6

Continuous Region

Continuous Region:

a subset of a basic region that contains only one UN and one SN and defines the types of feasible continuous distillation products that are possible for an initial composition within this region

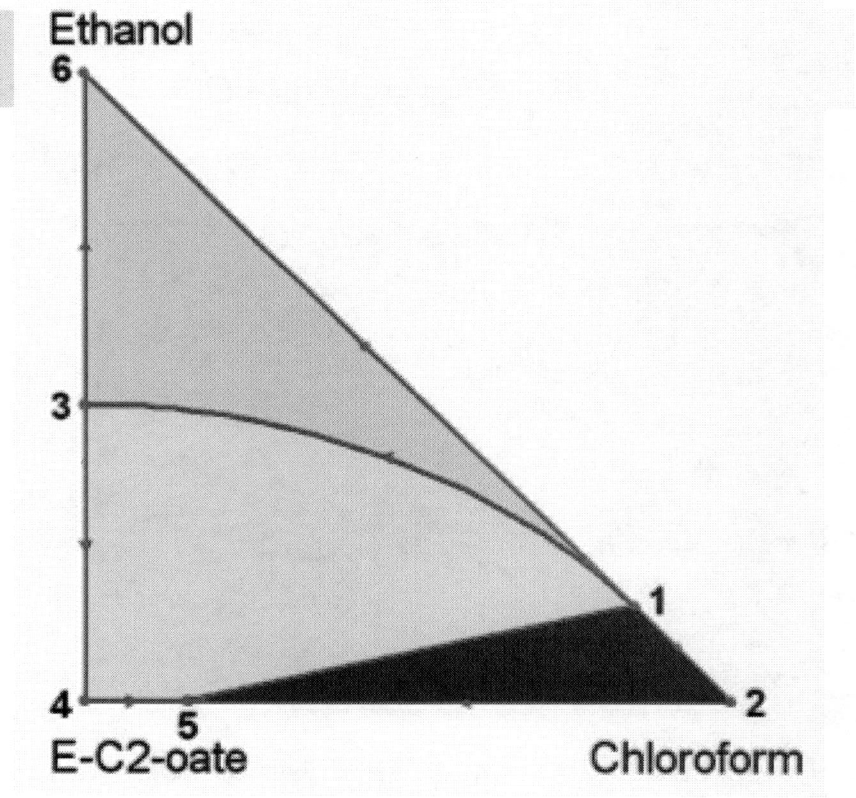

Continuous Distillation Regions:

101: 1 2 5
102: 1 3 4 5
201: 1 3 6

Batch Rectifier Region

Batch Rectifier Region:
a subset of a basic region that contains only one UN and defines the sequence of batch rectifier distillation product cuts that are possible for an initial composition within this region

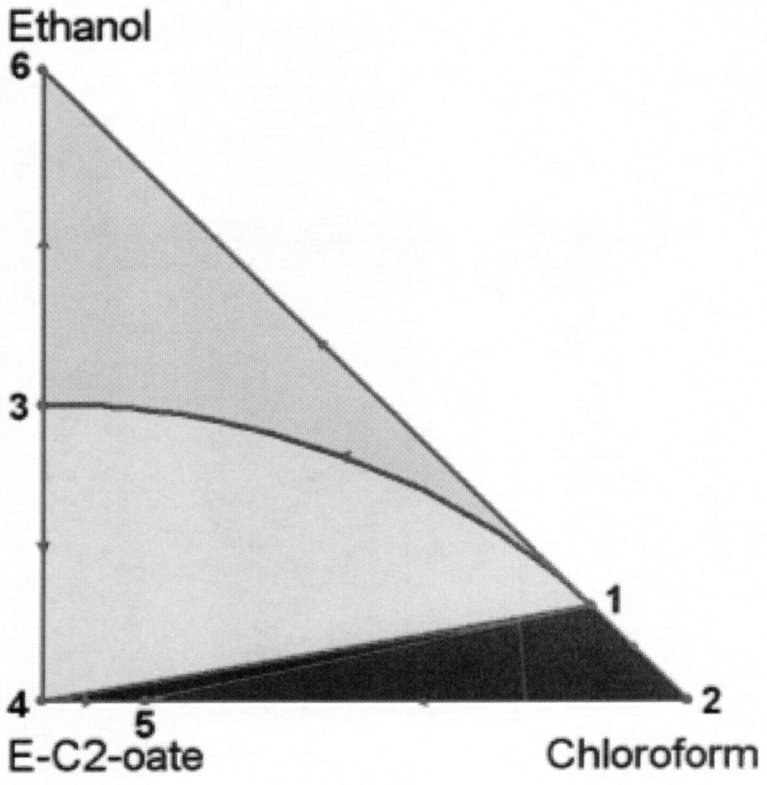

Batch Rectifier Distillation Region

10101: 1 2 5
10201: 1 3 4
10202: 1 4 5
20101: 1 3 6

Batch Stripper Region

Batch Stripper Region:
a subset of a basic region that contains only one SN and defines the sequence of batch stripper distillation product cuts that are possible for an initial composition within this region

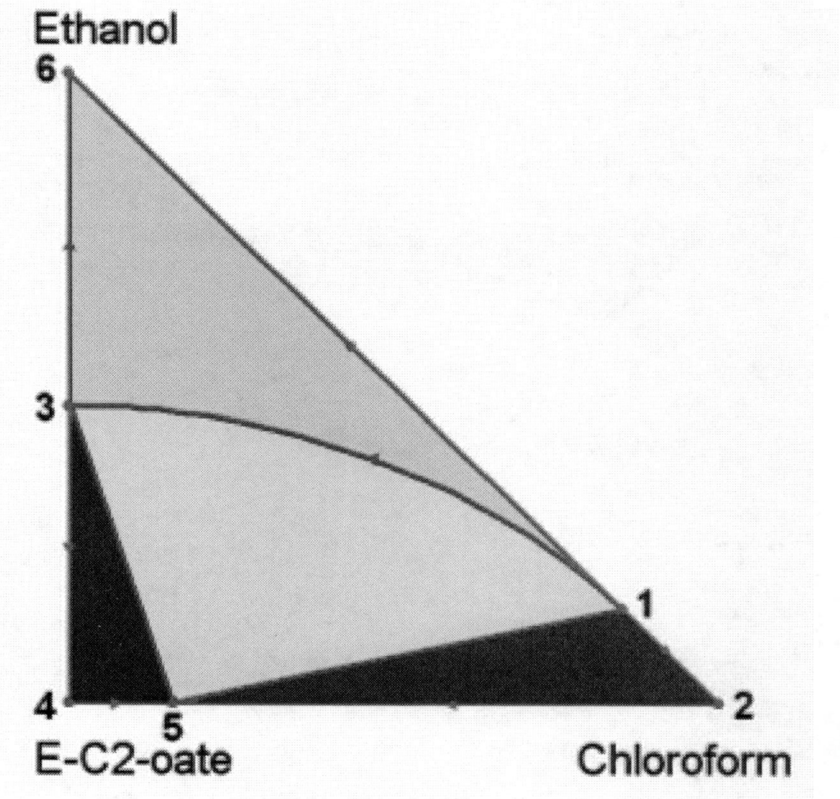

Batch Stripper Distillation Regions:
10101: 1 2 5
10201: 3 4 5
10202: 1 3 5
20101: 1 3 6

Distillation Regions

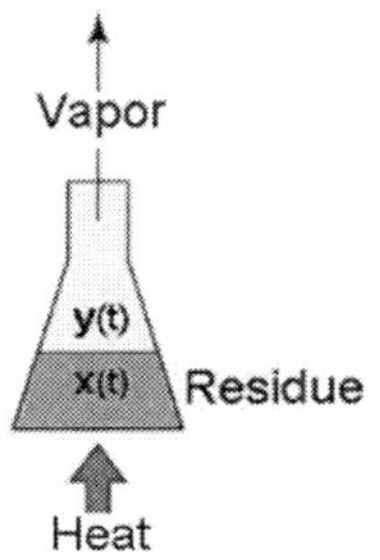

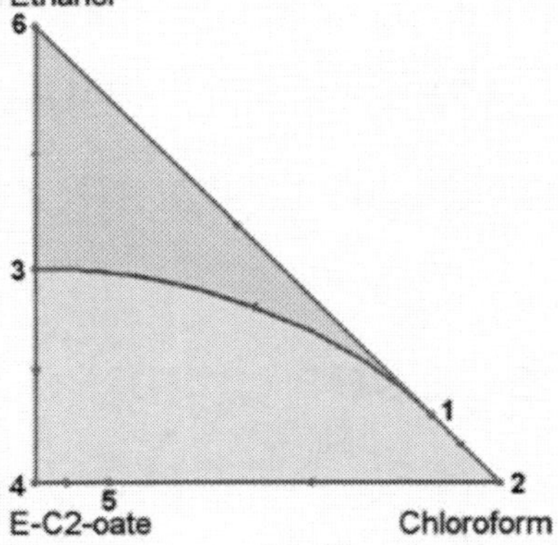

Basic Distillation Regions

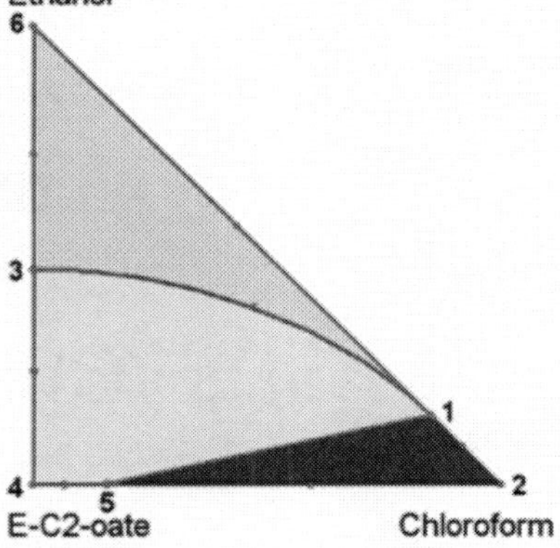

Continuous Distillation Regions

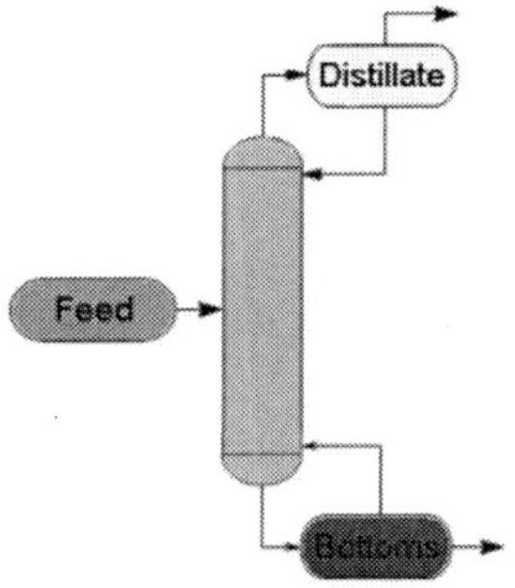

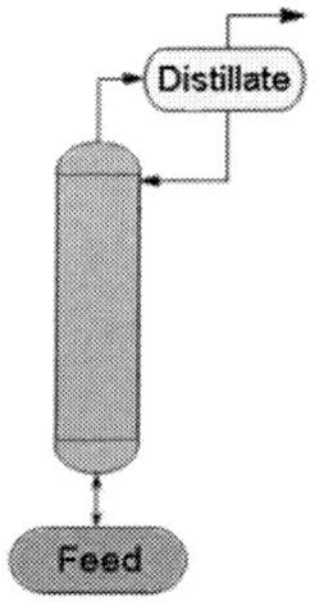

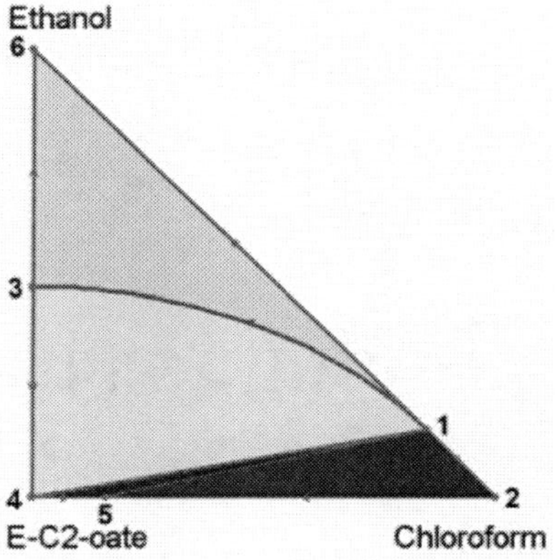

Batch Rectifier Distillation Regions

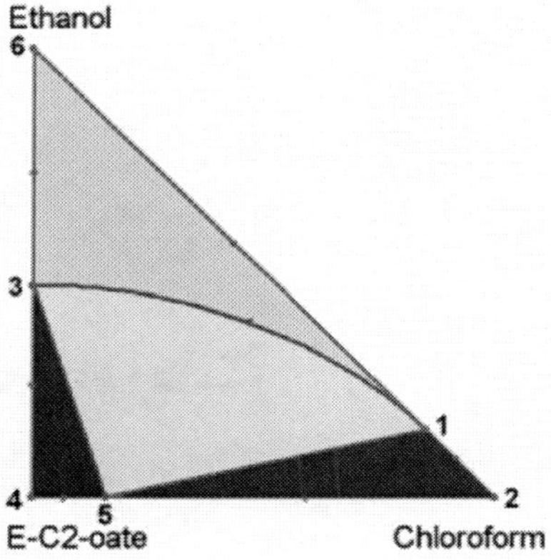

Batch Stripper Distillation Regions

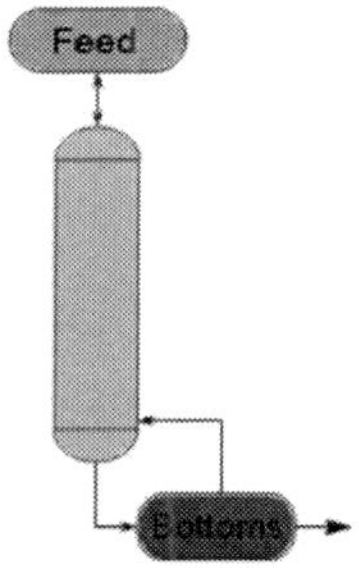

Generation of Distillation Regions

- Prepare a rigorous thermodynamic model of the mixture, estimate and regress parameters from experimental data[1]
- Calculate all azeotropes[2,3]
- Calculate adjacency and reachability matrices[4]
- Find all basic distillation regions[4]
- Find all distinct paths in each basic region (depth-first search algorithm[5])
- Find all continuous distillation regions
- Divide each continuous distillation region into batch distillation regions for batch rectifier/stripper

1 Wasylkiewicz, S. K., L. C. Kobylka, and M. A. Satyro, *Chemical Engineering*, 80-85, August 1999.
2 Fidkowski, Z. T.; Malone, M. F.; Doherty, M. F. *Comput. Chem. Eng.* 1993, *17*, 1141-1155.
3 Wasylkiewicz, S. K., M. F. Doherty, and M. F. Malone, *Ind. Eng. Chem. Res.*, 38, 4901-4912 (1999).
4 Rooks, R. E., V. Julka, M. F. Doherty, and M. F. Malone; *AIChE J.* (1998).
5 Gobbson, Algorithmic graph theory, Cambridge, 1987

Example: Structure of Separation Space

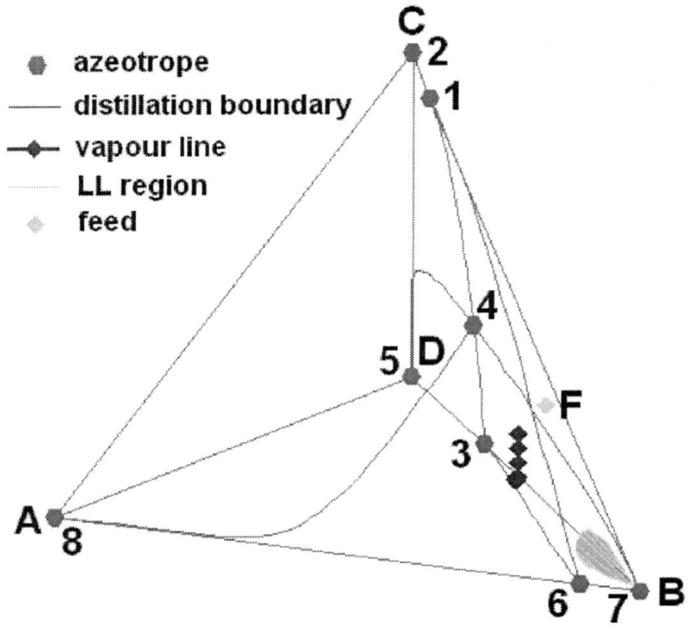

Adjacency matrix

	1	2	3	4	5	6	7	8
1	0	1	0	1	0	1	1	0
2	0	0	0	0	1	0	0	1
3	0	0	0	1	1	1	1	0
4	0	0	0	0	1	0	1	1
5	0	0	0	0	0	0	0	1
6	0	0	0	0	0	0	1	1
7	0	0	0	0	0	0	0	0
8	0	0	0	0	0	0	0	0

Reachability matrix

	1	2	3	4	5	6	7	8
1	1	1	0	1	1	1	1	1
2	0	1	0	0	1	0	0	1
3	0	0	1	1	1	1	1	1
4	0	0	0	1	1	0	1	1
5	0	0	0	0	1	0	0	1
6	0	0	0	0	0	1	1	1
7	0	0	0	0	0	0	1	0
8	0	0	0	0	0	0	0	1

Example: Distillation Regions

Basic Regions	Continuos Regions	Batch Rectifier/Stripper Regions
1: 1 4 6 7	**101**: 1 4 6 7	**10101**: 1 4 6 7
2: 1 2 4 5 6 8	**201**: 1 2 5 8	**20101**: 1 2 5 8
	202: 1 4 5 8	**20201**: 1 4 5 8
	203: 1 4 6 8	**20301**: 1 4 6 8
3: 3 4 6 7	**301**: 3 4 6 7	**30101**: 3 4 6 7
4: 3 4 5 6 8	**401**: 3 4 5 8	**40101**: 3 4 5 8
	402: 3 4 6 8	**40201**: 3 4 6 8

Primary Recycles

A primary recycle is detected when the composition of a product stream from the last added separator:

- is equal to the composition of a feed stream already present in the partially synthesized sequence (***first kind***);
- is not equal to the composition of the existing unit feed although if added to the unit as a second feed will only change flow rates but not compositions of the products (***second kind***);
- is not equal to the composition of the existing unit feed but both streams have the same components and addition of the stream to the unit will not move the unit from the original distillation region (***third kind***).

Secondary Recycles

- A *secondary recycle* stream is introduced in order to shift the total feed composition and by doing this change the functionality of a separation unit.
- In an ***internal secondary recycle*** a species present in the sequence feed is introduced as the separating agent. It can be a pure component produced somewhere downstream in the sequence or any other intermediate stream.
- In an ***external secondary recycle*** an external species is introduced as the separating agent.

- Recycles reduce the number of separation steps.
- Units that perform identical tasks are merged by recycling.
- DISTIL detects primary recycle streams and their destinations

Synthesis of Separation Sequences

Synthesis without secondary recycles

- Find all continuous/batch distillation regions
- Find the continuous/batch distillation region the feed is in
- Determine sequence of splits/cuts for the given feed composition
- Select primary recycles and their destinations

Synthesis with secondary recycles

- Find the preferred distillation region (where the best recovery of desired product can be accomplished)
- Find the optimal separation/mixing sequence

Example: Generalized Pressure Swing Distillation

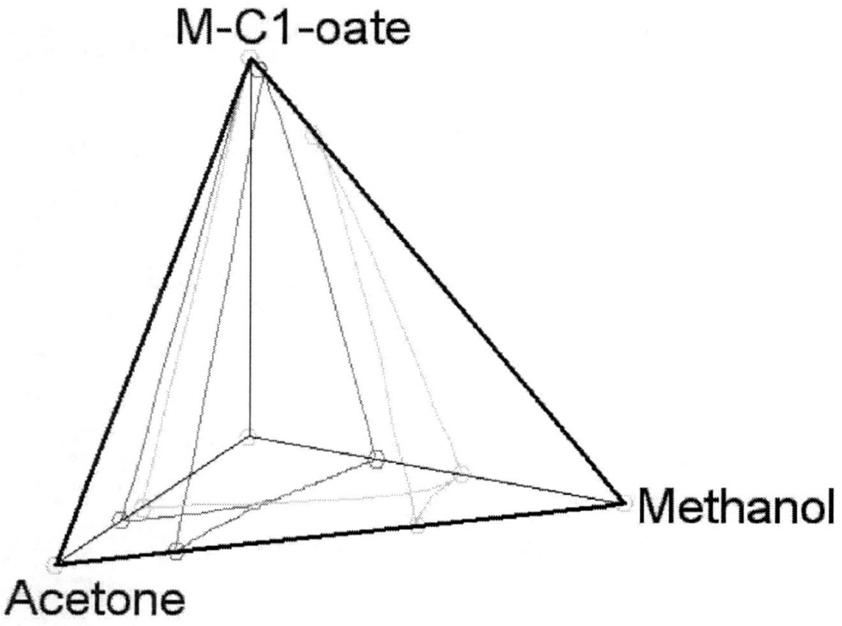

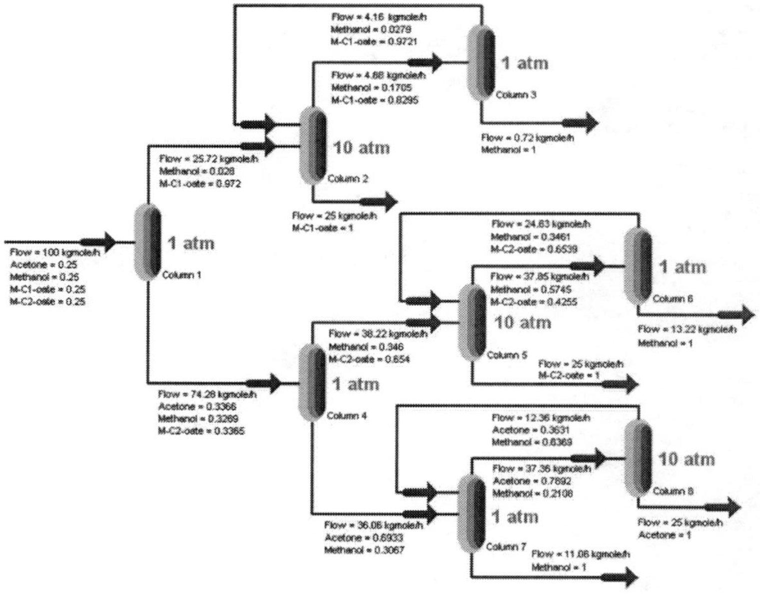

Azeotropes, distillation regions and distillation boundaries (left) and a sequence of distillation columns (right) for separation of an equimolar mixture of acetone – methanol – methyl formate – methyl acetate at pressures 1 and 10 atm

Example: Heterogeneous Azeotropic Distillation

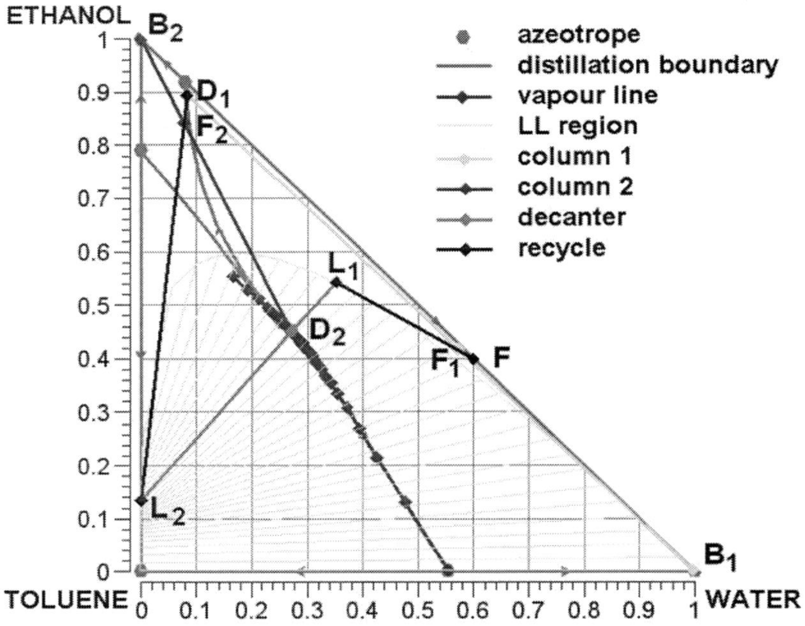

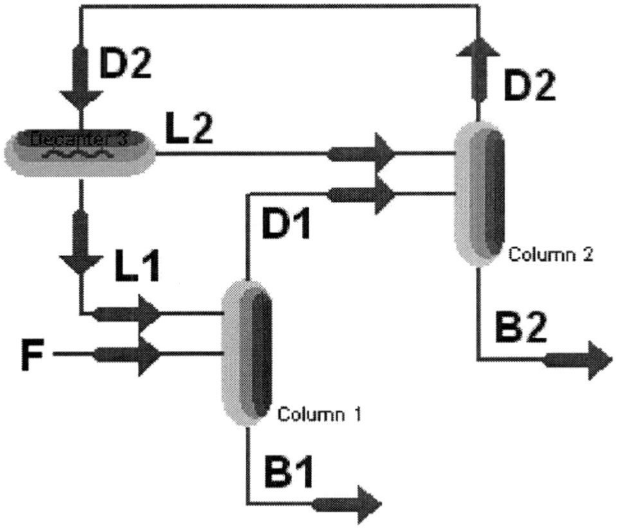

Azeotropes, LL region, distillation regions, distillation boundaries and mass balances (left) and the heterogeneous azeotropic distillation sequence for the separation of ethanol – water azeotrope using toluene as an entrainer

Example: Generalized Heterogeneous Azeotropic Distillation

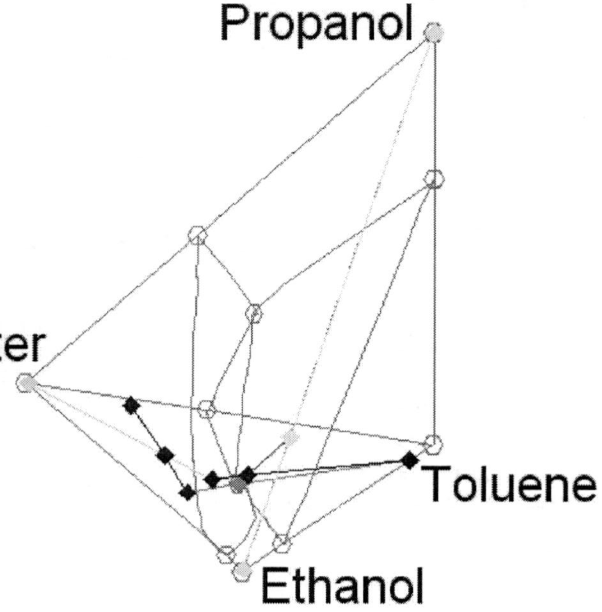

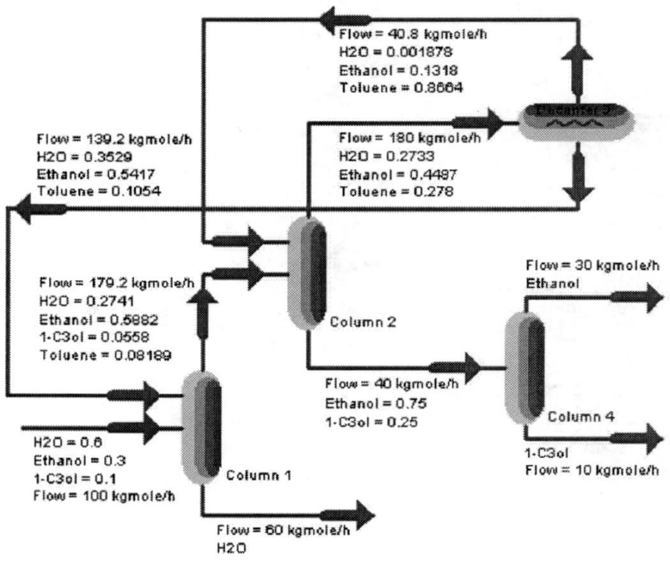

The heterogeneous azeotropic distillation process for separation of ethanol-propanol-water mixture using toluene as an entrainer

Example: Generalized Heterogeneous Azeotropic Distillation

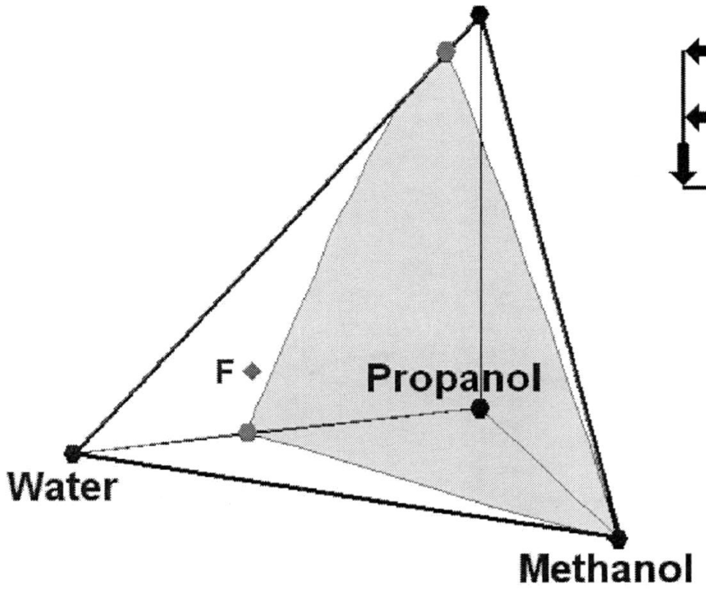

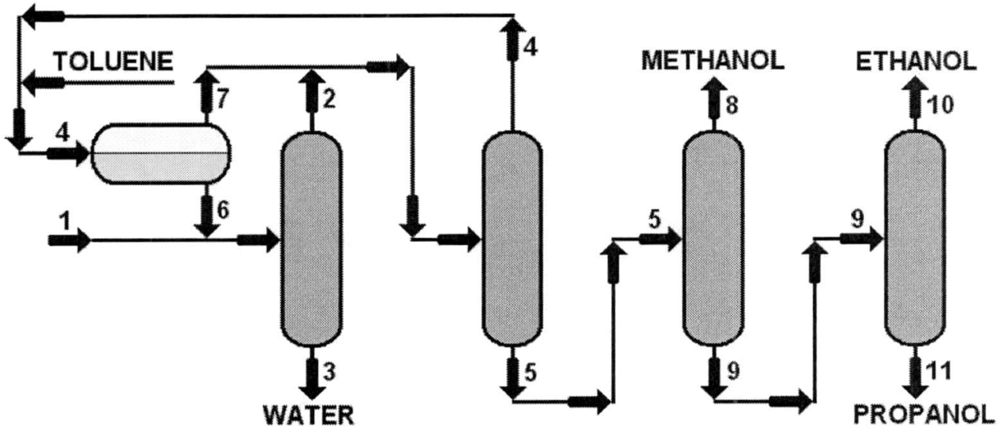

Two azeotropes and one distillation boundary in a mixture of water, methanol, ethanol and propanol (left) and a sequence of four distillation columns and one decanter (right) for separation of the mixture using toluene as an entrainer

Example: Internal Secondary Recycle

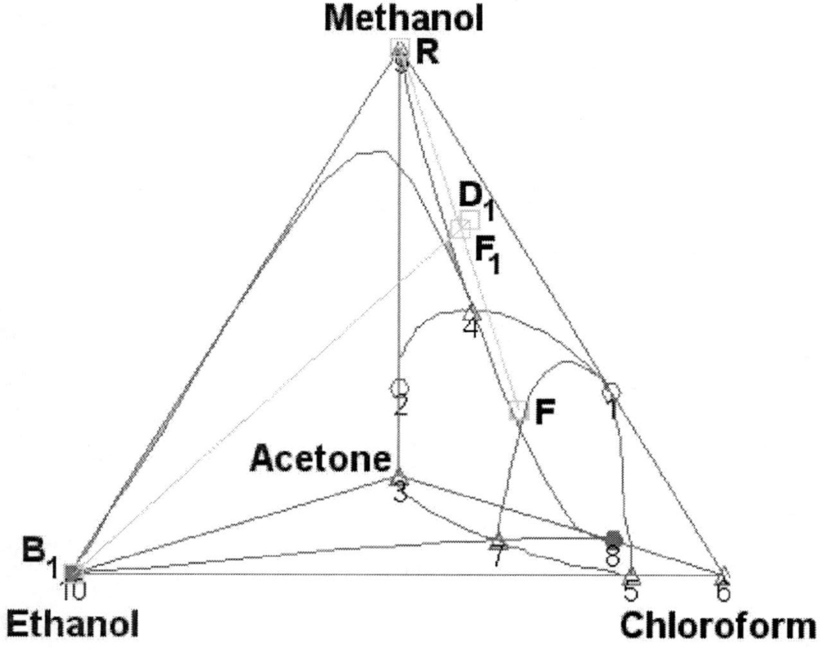

Distillation Regions	Temperature [C]	Name
Region 1	53.82	Chloroform - Methanol
	57.32	Chloroform - Methanol - Acetone
	59.62	Ethanol - Chloroform
	61.10	Chloroform
	63.18	Ethanol - Chloroform - Acetone
	64.45	Chloroform - Acetone
Region 2	53.82	Chloroform - Methanol
	57.32	Chloroform - Methanol - Acetone
	59.62	Ethanol - Chloroform
	63.18	Ethanol - Chloroform - Acetone
	64.48	Methanol
	78.17	Ethanol
Region 3	55.35	Methanol - Acetone
	56.06	Acetone
	57.32	Chloroform - Methanol - Acetone
	63.18	Ethanol - Chloroform - Acetone
	64.45	Chloroform - Acetone
Region 4	55.35	Methanol - Acetone
	56.06	Acetone
	57.32	Chloroform - Methanol - Acetone
	63.18	Ethanol - Chloroform - Acetone
	64.48	Methanol
	78.17	Ethanol

Composition space (left) and distillation regions (right) for the mixture of ethanol, chloroform, methanol and acetone at 1 atm. Mass balances for the first column and the recycle are also shown.

Example: Internal Secondary Recycle

- None of the feasible splits for the original feed ***F*** fulfills our objective in the synthesis: complete recovery of ethanol from the mixture and recycle all other components to the process.
- Feed stream ***F*** is in Region 1. It is impossible to obtain ethanol as a pure product in a simple distillation column placed in this region.
- Regions 2 and 4 contain ethanol as stable nodes ("*preferred*" regions).
- Add a recycle stream ***R*** to the stream ***F*** to obtain the overall feed F_1 in Region 2 or 3.
- Candidate for the recycle stream: a pure component or an azeotrope that can be produced from Region 2 or 3 after complete recovery of ethanol.
- Methanol is a good choice. After complete recovery of ethanol in the bottoms we have to use distillate from the first column to produce sufficient stream of methanol for the recycle. This can be done in the second distillation column of the sequence operating at lower pressure.

Example: Internal Secondary Recycle

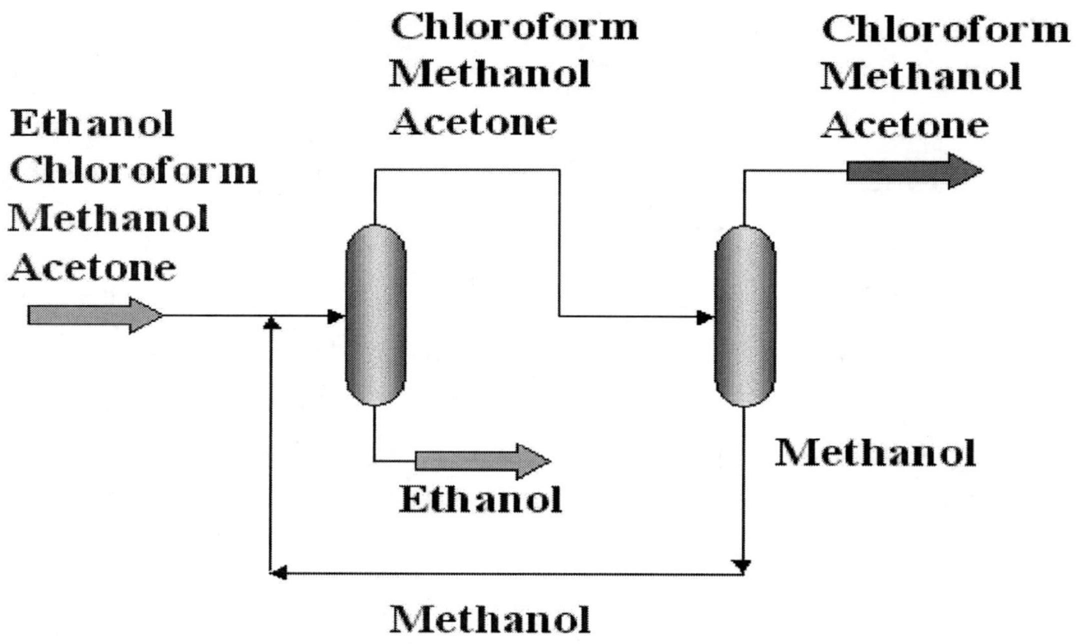

Sequence of distillation columns with internal secondary recycle

Conclusions

- A new algorithm for calculation of continuous/batch distillation regions has been developed based on information from adjacency and reachability matrices.

- Based on information about distillation regions and distillation boundaries, a new algorithm for synthesis of continuous/batch distillation sequences has been developed. It facilitates efficient crossing distillation boundaries and quick finding “preferred” distillation regions for particular separation task. It can be used for mixtures of any number of components.

83f A General Rate-Based Model for Industrial Reactive Absorption and Desorption Processes in Sour Gas Treatment

Berthold Diekjakobs, Günter Wozny, Holger Thielert, Jens-Uwe Repke, Robin Thiele

Increased demand on the global coke market led to a nearly eightfold price for coke. Building new plants with a modern competitive design for the gas section is therefore, also due to the overaged cokery plants in the world and increasingly stringent environmental regulations, imminent. In the industry design of gas purification plants is still often done using empirical methods. Gas purification using chemical absorption is a well established process. Alternative approaches like hybrid processes including membrane technology are still not economically feasible. Therefore, in order to advance gas purification processes rigorous models are required in order to improve existing chemical absorption processes.

In this project reactive absorption and desorption processes are rigorously modeled using thermodynamic and chemical engineering fundamentals. Frequently installed processes like ammonia-hydrogen sulfide scrubbing processes and amine-based processes have already been modeled with great success. However, models for processes using potash for coke-oven-gas purification, like the promising vacasulf process, are still not available. During the start-up of a recently built industrial vacasulf plant in Indianapolis, the experimental results showed a surprisingly large increase in efficiency if the alkaline concentration is changed compared to the initial design, which was not predicted by the empirical methods used in the design stage.

Thus, the vacasulf process was rigorously modeled for packed towers using a rate-based approach including the components H_2S, CO_2, NH_3, HCN, H_2O, KOH as well as inert components in the raw gas stream and salts in the liquid phase. Heat and mass transfer is modeled by the film theory. The enhancement of mass transfer by chemical reactions is taken into account using appropriate enhancement factor correlations. Dissociation of the different components in the electrolyte mixture into 16 species is taken into account using activity coefficients. Heat effects during absorption and desorption are accounted for by inclusion of standard enthalpies of formation. For fast and robust convergence the model has been implemented in FORTRAN to allow the application of customized convergence strategies. The model was then fully integrated into a commercial flowsheet simulator and is now being used in industrial practice for designing new plants.

Furthermore, the contribution also shows the validation of the novel model by own pilot plant experiments with all relevant components (except HCN) for both absorption and desorption in a pilot plant under lab conditions and additionally by industrial on-site measurements, which were taken during extensive examinations at the aforementioned vacasulf plant. The experimental data compare well with the simulation results. The model is then used to find optimum operating parameters, i.e. pressure range for the regeneration section, alkaline concentration range for optimal efficiency and selectivity of the absorption section and the flow rate of the circulating absorbent. The simultaneous adjustment of design parameters based on the newly developed process model facilitates a systematic improvement of the industrial process. In the lecture the model, the experimental facilities and the results will be described in detail.

83g Heterogeneously Catalyzed Synthesis of n-Butyl Acetate: Reaction Kinetics, Side Reactions and Application to Reactive Distillation

Hans Hasse, Oliver Bailer, Sandra Parada, Sergej Blagov

Reactive distillation is an attractive process for the production of many esters. However, in this highly integrated process side reactions may occur which do not play a role in conventional process. The question of the migitation of these side reactions often decides if reactive distillation can be used in practice. This topic is addressed in the present work by a combination of experimental, modeling and simulation studies using the heterogeneously catalyzed synthesis of n-butyl acetate from n-butanol and acetic acid as an example.

Reaction kinetics of the main and the side reactions in that system were investigated in an isothermal fixed bed tubular reactor at temperatures between 80 - 130 C using three ionexchange resins, which differ in their sulfonation. In Purolite CT 269 and Amberlyst 48, the micro particles forming the catalyst pellets are fully sulfonated, whereas in Amberlyst 46 only the surface of the micro particles is sulfonated. Therefore, the ion exchange capacities of Purolite CT 269 and Amberlyst 48 are similar whereas that of Amberlyst 46 is five times lower. The observed side products are isomers of butenes, di-n-butyl ether, sec.-butyl-n-butyl ether as well as sec.-butanol and sec.-butyl acetate. For Purolite CT 269 and Amberlyst 48, the side reactions proceed at similar rates. For these two catalysts, the side reactions are relatively fast and significant amounts of side products are produced. The results for the surface-sulfonated Amberlyst 46 are different: except small quantities of di-n-butyl ether, no side products were found. Despite the lower ion-exchange capacity of Amberlyst 46, the reaction rate of the esterification to n-butyl acetate is almost equal for all three catalysts. This leads to the conclusion that the relatively fast esterification takes place primarily on the surface of the micro particles whereas the significantly slower side reactions mainly occur inside these particles. Therefore, surface-sulfonated catalysts are very promising for esterifications in reactive distillation columns, as they allow maintaining high rates of the main reaction and simultaneously avoid side reactions.

To investigate how the reaction kinetics determined in the fixed bed reactor can be applied for reactive distillation packings, experiments were carried out in a trickle bed reactor (d = 50 mm) that was packed with the structured catalytic packings Katapak-S and Katapak-SP11 (h = 1.5 m). The comparison of the reaction kinetic data from the fixed and the trickle bed reactor shows that both the rates of the esterification and the side reactions are lower in the trickle bed than in the fixed bed. This can be explained by the differences in fluid dynamic regimes in the two reactors. The differences can easily be taken into account in the reaction kinetic model: the model developed on the basis of the fixed bed experiments can successfully be used for the trickle bed experiments when the reaction kinetic constants are lowered by about 20 %. This is done using a transfer factor that has the same number for all experiments.

Finally, the application of the obtained reaction kinetic model both for the main and the side reactions to reactive distillation experiments is discussed. For that purpose, available experimental data from measurements in a reactive distillation column (d = 81 mm) equipped with Katapak-S (h = 1 ÷ 2 m) are used. The simulations are based on a stage-model, assuming physical equilibrium between vapor and liquid on each stage, taking liquid phase reaction kinetics into account. The model results are compared to the experimental data from the reactive distillation column.

83h Solubility of Formaldehyde and Trioxane in Aqueous Solutions

Hans Hasse, Thomas Grutzner

Formaldehyde is one of the most important chemical intermediates.

84a Push Valve Experience on Distillation Trays

Daniel R Summers

Sulzer Chemtech has developed and employed push valves on all types of trays for thepast several years. Numerous towers have employed this technology and the feedback from its use is outstanding. A discussion of this technology, its development, application and operational feedback will be presented.

84b Control of Liquid Flow on Sieve Trays

G X Chen, Tony Cai

The understanding of liquid movement on a tray is essential for the better prediction of tray efficiency and hydraulic performance. Liquid flow non-idealities on single-pass cross-flow sieve trays were studied in a 4-ft diameter simulator using colored dye technique. A practical and simple solution was found to remove these non-idealities. A shaped outlet weir, together with a shaped and slotted downcomer baffle and liquid guide-vanes, produced a uniform flow across the tray. With these devices, no retrograde flow was observed. The design was tested over a wide range of flow rates from 1 to 8 gpm/in of weir. Experimental results and flow profiles are presented and discussed.

84c Tips for the Successful Design and Installation of Cartridge Trays

Ruth R Sands

Cartridge trays can be the perfect solution for distillation columns that are less than three foot in diameter and that have a unique need for trays. Systems that foul or require a large turndown can be accommodated with various tray designs installed as a cartridge unit. DuPont has many systems that fall into this category. This paper will present good practices for both the design and installation phases of projects that lead to successful installations and start-ups.

84d Modeling the Minimum Entrainment on Sieve Tray

Shuzo Ohe

Distillation column efficiency and capacity are both affected by the entrainment on a sieve tray. The entrainment increases with the increase of vapor load at constant liquid load. However, at constant vapor load, entrainment behaves somewhat differently: it has a minimum value. This phenomenon was uncovered by systematic study on a wide range of trays and systems at FRI (Fractionation Research Incorporated). The dominant explanation offered for this seems to be that the minimum is caused by the transition of vapor liquid contact from spray to bubbly regime on the tray. We, however, argue from an entirely different point of view. We assume that the longer the residence time of the liquid on the tray, the greater the chances of generation of entrainment. From data released by FRI, we ascertained a new relationship: that between minimum entrainment and minimum residence time at constant vapor rate and increasing liquid rate for sieve trays. The liquid load at which the entrainment becomes minimum at constant vapor load operation coincides with the liquid load at which the residence time of liquid on the tray becomes minimum. In our new approach to residence time, we define volume as the product of the bubbling area and the observed clear liquid height.

If the clear liquid height is known, the relation between it and liquid load can be applied to predict the minimum entrainment. A series of experiments for an FRI pressure drop study provided us clear liquid height values as well. This afforded us an opportunity to predict the minimum of entrainment from the value of clear liquid height alone. Clear liquid height has an important role in determining the residence time, so its relation to liquid and vapor loads should be expressed clearly. In searching for such clarity, we examined conventional Francis-type correlation equations for calculating the clear liquid height. However, we found that, using these equations, the obtained shape of the change in clear liquid height against liquid load is quite different from that of the observed clear liquid height. This difference in shape of clear liquid height is quite noticeably marked when 2/3 is used as the exponent of liquid load: Q. The observed value appears in graphs as a concave curve, while the calculated value is a convex curve. Such results suggest that correlation equations using 2/3 as the exponent are inadequate to fully express the change of clear liquid height against liquid load. Thus, we must look for a new way to consider this correlation of clear liquid height with liquid load. In the absence of highly reliable and realistic equations for calculating clear liquid height, the need is even greater for an accurate and easy-to-use empirical method.

84e Advances in Styrene Fractionation with INTALOX® Packed Tower Systems: Part 1 - Column Internals

Christoph Ender, Gail Hausch, Randall A Sommerfeldt

Structured packing has been used commercially in styrene service for over two decades. New design advances continue to be developed to meet ever-increasing process demands and provide superior tower performance. INTALOX® Packed Tower Systems by Koch-Glitsch combine high performance FLEXIPAC® HC® Structured Packing with state-of-the-art high performance internals tailored to meet specific process objectives. Driven by recent styrene projects, technology improvements have been developed and implemented.

Three major areas of improvements and innovations were made in column internals design during a recent revamp project of a world-scale styrene splitter column:

1. Liquid distribution technology included the Model 156 Trough Distributor with Enhanced Baffles for optimum liquid distribution to structured packing in low to moderate liquid load applications. For quality control and design validation, a 12 m [40 ft] diameter indoor distributor test facility was constructed to water flow test distributors in complete, full diameter assembly.

2. Innovative design improvements were developed and patented for large diameter liquid collectors to optimize liquid cross mixing while minimizing the height required between packed beds and eliminating the need for mixing drums.

3. CFD (Computational Fluid Dynamics) studies were used to advance vapor distribution technology in this critical service application. In this project, CFD was used to optimize the vapor distribution hardware, including nozzle deflectors, to reduce complexity and minimize pressure drop of the vapor distributor tray.

84f Design Guidelines for Distillation Columns in Ethyl-benzene and Styrene Monomer Service

Andrew Sloley, Karl Kolmentz, Krishnamoorthy Senthil, Mike Hightower, Tau Yee Lim, Timothy M Zygula, Wai Kiong Ng

Styrene monomer is the fourth largest chemical produced on an industrial scale and most ethyl-benzene is utilized in styrene monomer production. The largest chemical produced on an industrial scale is ammonia for fertilizer production, followed by crude oil refining, and then ethylene by furnace pyrolysis. Styrene monomer has been manufactured commercially for more than fifty years with advances in the key unit operation areas of reactor design and distillation.

The original distillation internals were trays, which have faced many challenges in services that are prone to polymerization. The higher pressure drop and longer residence time of trays are disadvantages in this service. First generation structured packing was introduced in 1980's. This advance greatly improved styrene distillation operations. The limit of first generation structured packing was capacity, where the stage efficiency began to decrease.

The rapid progress in computational fluid dynamic modeling (CFD) over the last ten years has also led to further improvements in the optimization of column internal design. CFD modeling of the structured packing, vapor-liquid distributors, and feed inlets can improve the performance of a distillation column.

Contemporary styrene unit designs include second generation structured packing and optimized liquid-vapor distributors. In the current designs, consideration of packing bed height and how the packing bed height affect stage efficiency due to mal-distribution should be reviewed. Designs are now limited by the maximum allowable pressure drop in order to avoid bottoms temperatures above 120 C (248 F) to 122 C (252 F). The authors will review the progression of Styrene Monomer Distillation advances and where the future research might be exploring.

84g Study of Intermediate Reboilers and Condensers for Distillation Columns for Binary Systems

Ashok V Naimpally, Martin Carlsson, Matt Keating, Osman Jance, Romit N Ketkar

An analytical study was done using both, an intermediate reboiler and an intermediate condenser in a continuous distillation system processing different binary mixtures. In the study, first an intermediate reboiler was used alone. Then a study was done using just the intermediate condenser. Lastly, the case where both the intermediate condenser and the intermediate reboiler were present simultaneously was also studied. Intermediate reboilers and intermediate condensers can reduce the energy cost and increase the chance of recovery and also improve the internal operation of the column.

In this study, an equation was derived along the lines of Smoker's equation to compute the number of stages. This was done for all 3 cases: intermediate reboiler alone; intermediate condenser alone and finally, both being present. For each case the column was divided into either 4 or 6 sections and then Smoker's equation was applied to each section separately and the total number of stages in that particular section was calculated. The total number of stages is found out by adding the number of stages in each section.

The next step in the study was to find the effect of the following 7 variables on the number of stages in the column: character of the feed; placement tray of intermediate reboiler; fraction of liquid vaporized in the intermediate reboiler; placement tray of intermediate condenser; fraction of vapor condensed in the intermediate condenser; the relative volatility of the mixture; reflux ratio. The results are shown in detail in graphical form. One of the surprising conclusions obtained is that for most mixture, the fractions of liquid reboiled had only a slight effect on the number of stages.

The last step in the study has been the derivation of a formula of the minimum reflux ratio and the minimum reboil ratio for the case of the column with intermediate reboiler and condenser. The equation has been derived along the lines of Underwood's equation.

* Author for correspondence: Professor, Dept of Chemical Engineering, Calif State University, Long Beach, Calif 90840. e-mail: avnaimpa@csulb.edu

Novel Packing Technologies and Their Application in China

SUN Jinsheng, Li Xingang, WANG Tao and ZHENG Yanmei
National Engineering Research Center for Distillation Technology, Tianjin 300072,China

Abstract: To give a brief introduction of the technical level and condition of packed column design in China, the authors include information from the structure of nevel packing and practical design of advanced liquid and gas distributors. Industrial examples are given in construction and revamp of ethylene and crude refinery units to support the new ideas and advantages of the mentioned special internals.

Introduction

Packed tower is an assortment of widely used vapor-liquid and liquid-liquid contacting equipment for mass transfer in chemical and petroleum refining industry, especially for distillation operations under atmospheric, low and middle pressure, as well as other two-phrase contacting processes (e.g. gas absorption and cooling). The unique geometric shape, regularizing gas and liquid routine, controlling wall flow and channeling, eliminating scale-up effect, and, in the mean time, supplying large gas-liquid contacting surface bestows structured packings high separation efficiency and low pressure drop. Breakthroughs of packed tower had been made since the beginning of 70' when corrugated sheet metal packing (Mellapac) was developed by Sulzer Brothers Ltd. With the continuous development of structured packings, the situation of trayed column as leading role of distillation equipment has been challenged. The maximum I.D. of column with structured packings is up to 14m, which breaks the traditional conception of the diameter limit of packed towers.

It is a prerequisite for ensuring the high efficiency of packed tower that the packings should be rationally selected and the internals matchingly designed. Internals in packed tower commonly include packing support unit, bed hold-down unit, liquid distributor, vapor distributor, etc.

The quality of liquid distributor design in packed towers directly affects the uniformity and distribution of fluid. Liquid maldistribution results in drop of separation ability of packed column, which restrains packed tower from its highest performance. Excellent liquid distributor ought to having the feature of uniform liquid distribution, large open area for vapor flow, low resistance, large turndown, less plugging, entrainment or foaming, easy manufacture and installation. Practice in recent years show that elaborate design and cautious installation of liquid distributor for guaranteeing proper initial distribution are the basic role in solving the problem of industrial scale-up of packed column.

Vapor distributor at gas inlet or reboiler return seems trivial in small columns, in which vapor can easily reach uniform distribution. But in towers with large diameter, especially in case of relative low bed height, vapor distribution is of extreme importance. Enough attention should be devoted to vapor maldistribution in column design, which leads to bypass of liquid and vapor flow in packed bed and severely decreases column separation efficiency. When gas-liquid mixture feed is involved, entrainment of liquid within gas phrase should also be taken into account in addition to assure even distribution of vapor and liquid.

When a new column is under design, the first to be considered is column type selection between the trayed and the packed. This step is decided by such various factors, instead of having unified standards, as designer's experience, knowledge, documentation and economic considerations, and so forth. Generally speaking, trayed tower is likely to be accepted by the firm and designer concerned because of being well-development and low price.

Among various trayed column categories, the efficiency discrepancy between each other is normally under 10% as long as correctly designed. Without special requirement, price, as the main aspect, should be given consideration to at the choice of trayed tower. Precise separation process requires large number of theoretical plates, while trayed tower solution usually needs much higher columns or a cascade of column, which is less adequate than the packed ones. In vacuum distillation for separation of high-boiling point or heat-sensitive mixtures, packed column is the most economic choice because of its low-pressure-drop nature. In brief, new towers under design require designers to choose proper type of column on the basis of process data and industrial practice in order to attain the aim of low cost and high profit.

To retrofit existing towers of fixed height and diameter for the purpose of increasing production and improving product quality, the best way is to transform all or part of their trayed sections into high-efficiency packing beds. Ordinarily, energy-saving can be obtained after revamp as pressure drop of columns with structured packing is much more reduced compared with previous trayed one.

It is noticeable that the advantage of low-pressure drop in packed column becomes unimportant in pressure distillation. The following facts may bring down the separation ability of a packed fractionator: tendency of non-uniform liquid flow and gas entrainment caused by high liquid flow and low vapor flow, liquid film thickening and viscosity increasing, etc.. Great care should be taken in case of pressure fractionator reformation by employing packed tower.

1. Structured Packings

1.1 Geometry of Corrugated Structured Packings

Structured packings in geometric pattern are regularly packed in columns. They permit realization of minimum pressure drop per theoretical stage, and therefore reduction of energy consumption by lowering column bottom temperature. They are most suitable for separating heat sensitive mixture. Additionally, they can eliminate random flow of fluid occurring in dumped packing beds and tend to uniform distribution.

The predecessors of most modern structured packings are gauze (Sulzer Packings) and sheet (Mellapac) metal corrugated types of Sulzer Brothers Ltd., after which came Flexipac by Koch Engineering company Inc. U.S.A., Gempak by Glitsch Inc. U.S.A., Montz packing by Montzs Company, Germany, and the like.

The above packings are all with corrugated shapes made from sheet metal of 0.2mm thick, in which there are punched orifices at the approximate porosity of 10%. The main effect of the orifices is to improve crosswise mixing of gas and uniform distribution of liquid. Grains and pits are stamped to the sheet metal to increase the irrigating feature of packing surface by liquid. Sheet metal is finally plaited to corrugated sheet at certain angles of inclination. The angle of corrugation vs. column axis is usually 45° (Y series) and 30° (X series). X series packings provides lower resistance and efficiency than their Y series. Prepared corrugated sheets are assembled to form cylindrical packing pies with alternating arrangement of adjacent corrugated sheets to make regular intersection of inclined channels, in which large gas-liquid contacting surface is provided for approaching uniform gas-liquid distribution and intimated contact. Neighboring packing pies are fitted close together in the column with their sheets cross at right angle to promote radical gaseous intermixing. Various models of structured packing are summarized in Table 1.

Table 1. Summary of various types of structured packings

Packing	*Company*	*Pattern description*
Mellapac	Sulzer Co., Switzerland, 70'	Integration of holed and corrugated sheets with stamped grained surface
Optiflow[1]	Sulzer Co., Switzerland, 1994	Integration of sheets with punched diamond geometry to induce optical flow
Flexipac[2]	Kock Inc. , U.S.A.	Same as Mellapac
Gempac	Glitsch Inc., U.S.A.,1982	Integration of corrugated sheets With holed and slit-orificed surface
Gempac-AW	Glitsch Inc., U.S.A.,1993	Integration of long-rectangular-holed corrugated sheets with slit-orificed surface
Intalox Structured Packing[3]	Norton Co., U.S.A.,1994	Integration of alternating corrugated metal-sheet
Rombopak	Kuhni Co.,Switzerland,1989	Integration of petal-like striped metal-sheets
MAX-PAC[4]	Jaeger Co.,U.S.A	With W openings in edges of corrugated sheets

Porous Corrugated Sheet Packing	Tianjin Univ.,P.R. China, 1985	Same as Mellapac
Corrugated Ring Packing	Qianlong Co. Tianjin, P.R. China, 1996	Integration of sheets with serial connection of stripes in both directions.

Figure 1 is the schematic diagram of geometrical dimension of corrugated sheets at their normal cross section. The t in the scheme represents thickness of corrugated sheets which should be as thin as possible to give light weight, low pressure drop and material economy. The h, height of corrugations, grows with specific surface area decreasing and efficiency lowering, while the flow channel hydraulic diameter enlarging, which contributes to lower resistance and arise throughput. The peak width, 2B, determined by the height of corrugations, h, and the angle between neighboring striped face, , usually valuing between 75~90° . The height of packing pie, H, normally scales from 100 to 300 mm. Small H zigzags the flow channel more frequently, which is advantageous to fluid redistribution, and at the same time, gives rise to resistance and difficulty in installation.

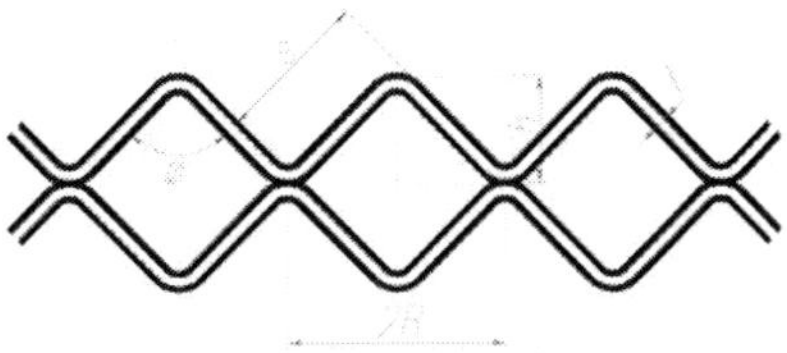

Figure 1 Geometry of corrugated sheets (cross section)

From the element enclosed by dotted line in Figure 1, equations are available for calculating specific surface area , porosity , hydraulic diameter D_h, and specific density $_p$ of packings:

Specific surface area

$$\alpha = \frac{2\left[S - t/\tan\left(\phi/2\right)\right]}{hB}(1-\sigma) \tag{1}$$

Porosity

$$\varepsilon = 1 - \frac{at}{2} \tag{2}$$

Specific density

$$\gamma_p = (1-\varepsilon)\gamma_m \tag{3}$$

Hydraulic diameter

$$D_h = \frac{4\varepsilon}{\alpha} \tag{4}$$

In above equations, . represents the porosity of corrugated sheets and $_m$ the density of sheet material.

1.2 Features of Corrugated Packings

Metal corrugated packings falls in two divisions of gauze and sheet packings that include several types respectively. Dimensional differences, such as specific surface area, hydraulic diameter and inclination of flow channels, indicate differences in performance, such as efficiency, capacity and pressure drop. The larger specific surface area and the more material consumed result in higher efficiency, and however, the higher pressure drop and the lower capacity.

At the beginning of 60', sheets in corrugated packings developed by Sulzer Brothers Ltd. were made of metal gauze including three main models of AX, BX and CY, among which BX and CY gauze packings were most commonly applied. Feature curves of BX and CY corrugated gauze packings[7] show that separation efficiency is increased and pressure drop decreased as vapor load going down. Besides, in case of high viscosity (> 10^{-3} Pa·s), pressure drop should be considered rising to some extend due to higher viscosity. CY packing possess twice as much theoretical stages per meter as BX, but having more pressure drop, lower capacity and higher cost.

High-performance gauze metal packings, also high in price, rely strictly on liquid distribution and are not suitable for handling materials with precipitation or solid matter. In early 70', Sulzer developed corrugated sheet metal packing with 12 models, in which 125Y(X), 250Y(X), 350Y(X) and 500Y(X) found their comparatively wide application with 250Y(X) the most. Similarly to corrugated gauze packings, the efficiency curves of corrugated sheet metal packings[5] indicate that, with vapor load going down, efficiency becomes somewhat higher and pressure drop lower. Operating turndown ratio of these packings is extremely large, and the load of vapor and liquid ranging in wide extent will not affect their separation efficiency.

1.3 ZUPAC[6, 7]

ZUPAC is a new generation of structured packing developed on the basis of corrugated sheet metal packing and Intalox random packing, with an alternative of the former, as shown in Figure 2.

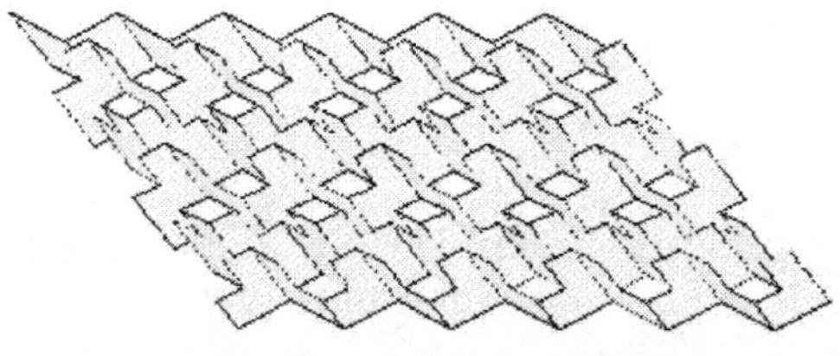

Figure 2 Outlook of ZUPAC sheet

Because of the corrugated openings in sheets bulging in opposite direction alternatively instead of punched orifices, surface area of this packing is 8 ~ 10% larger than that of corrugated sheet metal packing. The features of this packing are large surface area for mass transfer, optimized routine for vapor and liquid flow and intensified cross-mixing. Moreover, it is enormously superior to corrugated sheet metal packing in such aspects as anti-plugging capability, rigidity, pressure drop and therefore throughput.

2. Trough Plate Gas-Liquid Distributor (TPGLD)[8]

Trough Plate Gas-Liquid Distributor (TPGLD), as patented technology of Tianjin University, is an improvement to chimney tray distributor with some advantages of trough distributor. TPGLD, whose construction is illustrated in Figure 3, consists of gas risers and their covers, liquid discharge unit and a deck,. The major difference between TPGLD and chimney tray distributor is that two rows of orifices, upper ones larger and lower ones smaller, positioned at the top and middle part of the gas risers from which liquid is guided into low vapor velocity area by special discharge units—the drip guiders. This distributor has functions of liquid collecting, gas distribution, liquid distribution as well as side-drawing, and the following advantages other than uniform gas and liquid distribution:

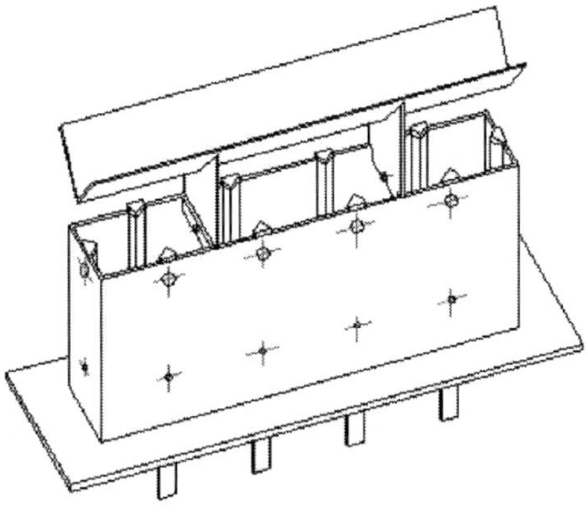

Figure 3 TPGLD unit

- Low height requirement: being a 3-in-1 gas-liquid distributor, its height requirement is 0.5 ~ 1 m lower than that of tube, spray nozzle and multi-stage trough distributor.
- Plugging resistance: heavy dirt in liquid layer sink to deck while the light ones float at liquid level, so middle-positioned orifices under liquid level is uneasy to be plugged.
- Anti-entrainment: gas and liquid flowing in separate path, entrainment does not appear at F factor in gas risers varying form 7 to 15 $[(m/s) \cdot (kg/m^3)^{0.5}]$.
- High turndown ratio: a designed turndown higher than 10:1 can be achieved.
- Low pressure drop: with opening over 30%, its pressure drop is lower than dry-plate pressure drop of conventional trays.

TPGLD can be designed in welded construction or assembly parts that can be inserted through manhole and then fitted in the column.

In view of the above advantages, TPGLD has been used in more than 200 tower in recent years, 20 of which in large scale. At present, the largest column utilizing TPGLD is the Φ

6600 (mm) gasoline quench tower at Qilu Petrochemical, and a distributor with column I.D. up to Φ8400 (mm) for Maoming Petrochemical Inc. had been designed.

3. Axial-Radical Gas Distributor

With the development of new type and high-performance packings of large voidage and low pressure drop, as well as their application in the case of large diameter and low packed bed height, vapor distribution in packed column appears most important. For large vacuum tower in petroleum refinery unit and gasoline quench tower in ethylene unit, whose gas inlet entrains massive liquid, specially designed distributors are needed for gas inlet with the features of low pressure drop, uniform vapor distribution and low entrainment. Gas distributors in common use include porous radical tube, dampened radical tube, tangential sloped partial horn, tangential/radical vapor horn and radical channel with bi-directional dampers, in which vapor horn is generally accepted to be excellent in performance.

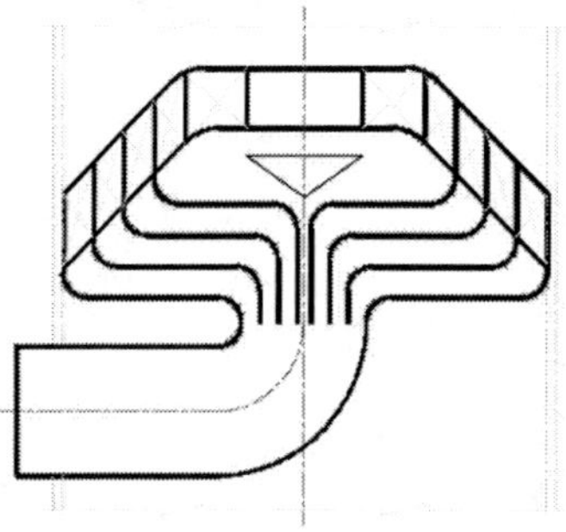

Figure 4 Axial-Radical Gas Distributor

Axial-radical gas distributor, as a new axial symmetrical, flow guiding and radically energy absorbing, is superior in performance to other plane symmetrical ones (vapor horn and radical channel with bi-directional dampers), Sketch of Axial-radical gas distributor can be seen in Figure 4. The main components of the distributor are gas inlet tube, radiator, flow guider and liquid catching and energy absorbing unit. High velocity gas flow with liquid entrainment enters the central radiator through inlet tube, where gas flow is split to radical flow in proportion. The radical flux is then changed to smooth axial flow in flow guilder. Flow guilder is equipped with liquid catching and energy absorbing unit to seize little liquid droplets in gas flow.

Axial-radical gas distributor features low pressure drop, uniform gas distribution and low entrainment. Testing results of this distributor in a laboratory column are summarized in Table 2 at liquid/gas ratio 1.2 and measured at 600mm above the distributor.

Table 2. Testing data of axial-radical distributor

F factor	Pressure Drop (Pa)	Degree of non-uniformity	Entrainment %
2.3	23	0.48	0.026
2.8	31	0.40	0.061
3.3	39	0.35	0.094

4. Industrial Applications

4.1 Revamp of Vacuum Distillation Unit in Petroleum Refinery

Since 70', technology of dry or semi-dry vacuum distillation has been rapidly progressed in China as a result of development of new packings and internals, lowering total pressure drop of columns and boosting vacuum in vaporization section to cut off energy consumption. So far, more than 20 units in China have been retrofitted to dry or slight wet vacuum distillation, considerable part of which with structured packings.

In May, 1990, with the aid of two patented technologies — combined liquid distributor (TPGLD) and aluminized C-steel corrugated sheet packing, Tianjin University, cooperated with SINOPEC Beijing Design Institute and Jinan Petroleum Refinery (JPR), successfully revamped fractionation section of netted trays between the third and the fourth side cut of JPR's lube vacuum tower. The tower became a fully packed one after the revamp. The retrofitting investment of RMB 499,000 Yuan was recovered in less than one month. After the reform, crude oil throughput increased from 1.2 Mt/y to 1.35 Mt/y, or 1.42 Mt/y in wet vacuum distillation case, so that the productivity of atmospheric and vacuum distillation unit going up by 2.28%, which resulted in surplus net profit of RMB 7 million Yuan per year. Comparative data of the tower before and after revamp is given in Table 3. In July 1992, with structured packing technology, Tianjin University revamped the 6400 mm fuel vacuum tower of Petroleum Refinery of Anqing Petrochemical Factory, which also succeeded.

With the changing of crude and oil product market, numbers of petroleum refineries have set out to transform their vacuum towers from fuel to lube or semi-lube ones for the sake of benefit. In conventional revamp of former trayed fuel oil vacuum fractionator, the height of column should be increased by 4~10m, which is not only uneconomical but also difficult in construction. Application of high-performence structured packing and new internal technology realizes a revamp without increasing the column height. For instance, conventional option of a fuel oil vacuum tower to be revamped of about 35m high means very difficult. However, tower height need not increase during the revamp by utilizing high-performance structured packing and patented TPGLD technology. And after the revamp, 5 side cut streams: #1 and #2 side cuts overlapped less than 10°C, #2 and #3

15°C, #3 and #4 30°C; distillate temperature (2% ~ 98% range) of #2 is 60°C, #3 55°C and #4 73°C.

Table 3. Comparison of Side Cut III and IV before and after revamp

Contents	*Unit*	*Before revamp (6th,June, 1987)*	*After revamp (7th,May, 1991)*
Tower I.D.	Mm	4200	4200
Tower Height	Mm	43100	43100
Fractionation Section Structure		4 Netted Trays	1.424 m JKB(sh)-250Y
Distributors			TPGLD
Crude Throughput	t/h	149.016	162.854
	t/a	$1.192X10^6$	$1.296X10^6$
Vacuum Tower Feed	t/h	99.606	119.491
VGO Production	t/h	43.261	48.485
Vacuum Residue	Ml	15	9
Distillate at 500°C			
VGO Productivity	%	29.03	29.77
Fractionation Section Pressure Drop	Kpa	0.933	0.467
#4 Side Cut Distillation Range	°C	74	47
#4 Side Cut Viscosity	m^2/s	14.15	11.29
#4 Side Cut Chromometry		6	4.5
#4 Side Cut Carbon Residue	w%	0.42	0.25
#3 & #4 Side Cut Overlap	°C	48	28
#4 & #5 Side Cut Overlap	°C	58	33

4.2 Revamp of Gasoline Quench Tower of Ethylene Unit

The 300,000 t/a ethylene unit of Qilu Petrochemical Inc. is a large scale plant introduced in the end of 70' from U.S.A. Gasoline quench tower is one of the most important separation equipment in ethylene cracking unit (see Figure 5). Pyrolysis gas is cooled to 210°C by quenching water in the quencher, and then enters the gasoline fractionator. In the gasoline fractionator, most of pyrolysis gas, steam and gasoline are cooled to below 108°C. Then, in water quench tower, the next step to separate pyrolysis gas, the mixture of rich gasoline and lean water is fed back partially to the gasoline fractionator as reflux and the rest is drawn out as product. Diesel and fuel oil in pyrolysis gas are drawn from the middle and the bottom parts of the fractionator respectively.

The production of ethylene unit of Qilu Petrochemical Inc. had only been able to reach to 80% of the designed capacity since May 1987, when it began to run. The dry point of gasoline was as high as 245°C and viscosity of quenching oil was 80~100 cm^2/s, greatly exceeding the designed. From 1989 to 1992, Qilu Petrochemical Inc., in cooperation with Beijing Petrochemical Engineering Inc., revamped extensively the unit to increase capacity of the tower to the expected and lower pressure drop. Nevertheless, dry point of gasoline and the viscosity of quench oil were still the bottleneck of the unit.

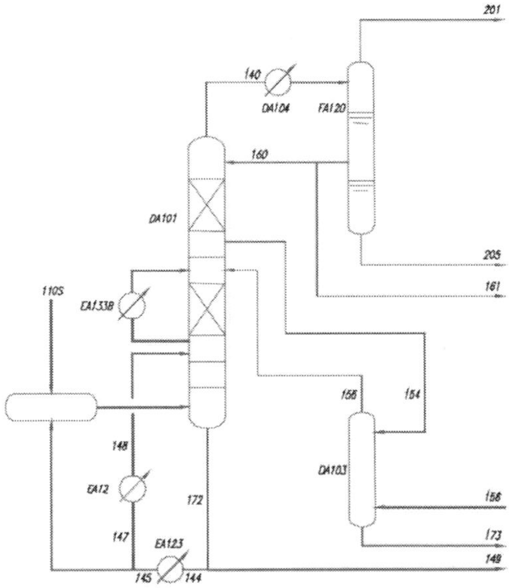

Figure 5 Scheme of Gasoline Quench Tower of Qilu Petrochem. Inc.

During the 1994 plant overhaul, Qilu Petrochemical Inc., together with Tianjin University, focused on the retrofitting of distributors in gasoline fractionator.

The distillation process in gasoline fractionator is a complicated one. Large quantities of enriched unsaturated aromatics in pyrolysis, such as styrene, indene and diethenyl benzene, etc., collect in the column, which are easy to cause coking by superheating. Once the coking happens, the orifices of the distributors are easily plugged. The original distributors were of orifice pan plate type with orifices opened in the plate so as to be facilely plugged. Practically, a great quantity of coke would be wept out in the overhaul each year, demonstrating that orifice pan plate distributors is not suitable in this case. Tianjin University replaced the orifice pan plate distributor with new TPGLD in the revamp with the aid of liquid self-distributed packed bed of 400 mm high to improve the liquid distribution. The production surpassed 115% with all designed target reached, and a satisfactory result of RMB 40,260,000 Yuan profit per year. Designed and operating data, as well as comparative conditions of consecutive two revamps are listed in Table 4.

Table 4. Comparison of Conditions of Tower DA-101 Before and After Revamp

Contents	*Designed Conditions*	*Tested Conditions Before Revamp*	*The 1st Revamp*	*The 2nd Revamp*
Top Temperature,C	108	118	110 ~115	108
Gasoline Dry Point,C	205	245	220 ~230	203 ~ 208
Pressure Drop,Kpa	12	12.5	9	5
Quench Oil Viscosity, cm^2/s	5	86	20 ~50	5

4.3 Revamp of FCCU Absorbing-stabilizing Unit

Catalytic cracking is a significant reprocessing of crude oil in petroleum refining industry. There are about 100 FCCUs in China. FCCU consists of three part, that is, reaction-regeneration system, fractionation system and absorbing-stabilizing system, whose main role is to transform heavy distillates like reduced crude, wax oil and deasphalted residuum to high quality dry gas, liquefied gas, gasoline and diesel under the simultaneous-action of heating and catalyst. Absorbing-stabilizing system contributes to separate fat gas and raw gasoline to dry gas, liquefied gas (C_3 & C_4 components) and debutanized gasoline with qualified vapor pressure.

In recent years, China's domestic petroleum refineries gradually carried out optimized reformation to reaction-regeneration system by new catalysts and process that outclass the gas productivity and gain considerable benefit. Simultaneously, they are further improving throughput of FCCU to get more profit. Thus, problems rise in FCCU not suiting to the increased load. Well or worse the FCCU absorbing-stabilizing system runs directly affects profit of the whole refinery. Normally the following problems commonly exist in the system:

- Dry gas unqualified. Owing to bad absorbing effect, bulk liquefied gas (C_3 components) containing in dry gas incurs loss of liquefied gas.
- High C_5 content in liquefied gas. Insufficient separation ability of stabilizer causes liquefied gas not up to standard for having entrainment of C_5 and lowering the gasoline yield.
- High C_4 content in gasoline. This results in disqualification of gasoline vapor pressure.
- Flooding in reabsorber leads to entrainment in dry gas.
- Large energy consumption. Over-desorption of desorption tower makes numerous C_3 & C_4 components recycling between the absorber and desorption tower, which increases energy consumption.

Due to the above problems in absorbing-stabilizing system and the requirement for production increase and fine chemical materials development, revamp to the system is urgently called for to improve its separation capability and capacity. Petroleum Refining Department of Chemical Engineering Research Center, Tianjin University has been working a lot at this point by successfully revamping several absorbing-stabilizing system giving the factories involved enormously high profit.[9]

The FCCU of Maoming Petrochemical Inc. was set up in 1972, with original designed capacity of 0.6 Mt/a. Through several revamps, its productivity has already exceeded 0.8 Mt/a. Aiming at enlarging capacity a step further, the stabilizer was retrofitted in December, 1992.[10]

The I.D. of rectifying section of Maoming's stabilizer is 2200 mm and the exhausting section 2800 mm. The liquid load in the latter is too large to operate normally before revamp. The retrofitting replaced all 20 trays in exhausting section by JKB-250Y corrugated sheet metal packing matched with two patented technologies—TPGLD and self distributed packing, leaving 20 floating valve trays in rectifying section unchanged. The TPGLD option guaranteed to retain packed height as high as 9.4 m. Whereas, a packed bed of 9.4 m high may result in liquid maldistribution in its lower part, to which the normal solution is liquid redistribution in the middle part of packed bed. Conventional redistribution is adding liquid collector and distributor between 2 packed beds by over 2m reduction of packed height. The liquid self redistributed packing used in this revamp can get the target of bettering liquid distribution without reducing bed height. After revamp, theoretical plate number in exhausting section increased greatly (by about 20) with production quality up to designed requirement and quantity up to the need of increasing capacity. See Table 5 for comparison between results before and after revamp.

Table 5. Comparison of Results Before and After Revamp

Contents	***June 11, 1991***	***Mar. 21, 1993***	***July 16, 1993***
Feed, t/hr	71.06	108.39	98.1
Gasoline, t/hr	59.98	92.42	84.95
LPG, t/hr	11.06	15.97	13.15
Reflux, m^3/hr	57	34	32
Top Temp.,°C	58.3	56	56
Bottom Temp.,°C	138.6	168	168
Top Pressure Drop, MPa	0.992	0.88	0.86

Table 5 shows that:

● Capacity of stabilizer rose. The value after revamp on July 16, 1996 was close to the designed of 103 t/hr, and on Mar. 21 reached 108.39 t/hr, surpassing the designed capacity by 5.2%.

● Reflux reduced. Before revamp the reflux was 57 m^3/hr, often as high as 80 m^3/hr or even higher. Reduction of reflux will further increase the throughput of stabilizer and realize FCCU energy economy.

● Stabilizer bottom temperature rose. This indicated lowering of vapor pressure of debutanized gasoline and increase of gasoline stabilizing depth. Bottom temperatures were 138.6°C under 0.992 MPa before revamp and 168°C under 0.86 ~ 0.88 MPa after.

Throughput of Zhanjiang Dongxing Petroleum Refinery's FCCU was 300,000 t/a with separate arrangement of absorber/desorption towers. Problems of bad absorbing effect, high atmospheric temperature as well as low compressor pressure, and the like, caused over 25% of liquid hydrocarbon burned at the flare. Gasoline components were often included in liquid hydrocarbon to affect its quality. Moreover, after the capacity of FCCU were expanded to 400,000 t/a, the above situation went worse. It can only be bettered by revamp.

The revamp was carried out with two column types—packed and trayed, whose main reconstruction is listed as follows:

● The original 1000 mm desorption column was changed to a completely packed absorber whose packed height was 12m, with trough distributor as top distributor, TPGLD as crude inlet distributor and trough distributor together with liquid collector as intermediate heat removing reflux distributor.

● The original 1200mm stabilizer was utilized as desorption column with structured packing in rectifying section and original valve trays remained with adjusted opening in exhausting section.

● A new stabilizer was erected of 1600 mm I.D. with 40 valve trays of 9.75% opening.

Comparison of before and after the revamp can be seen in Table 6.

It shows that after revamp, content of C_3 is increased and the heavier components is decreased which makes yield of liquid hydrocarbon becomes higher with significant profit. The total investment of the revamp was RMB 1,495,000 Yuan. Disregarding the increase in capacity, at least 30 tons of additional liquid hydrocarbon can be regained per day, meaning a yearly benefit of RMB 19,800,000 Yuan. The revamp investment can be returned in 25 days. Taking throughput ascent into account, the profit will be even more.

Operating pressure of Zhanjiang Dongxing Petroleum Refinery's absorbing-stabilizing system is 0.9 MPa (absolute pressure), which was somewhat lower. Raising the pressure to 1.1 ~ 1.2 MPa, the content of C_3 and the heavier in dry gas can be reduced to less than

1%, giving higher liquid hydrocarbon yield.

Table 6. Comparison of Performance Before and After Revamp

Date	*Throughput t/d*	*liquid hydrocarbon, t/d*	*liquid hydrocarbon yield, wt%*	*Content of C_3 and the heavier in dry gas, v%*
Before 29-May	872	64	7.34	25.6
Before 30-May	849	63	7.42	29.6
After 24-Aug.	819	104	12.7	1.2
After 25-Aug.	796	105	13.2	3.04
After 26-Aug.	789	95	12	3.86
After 28-Aug.	796	96	12.1	1.37
After 29-Aug.	817	106	13	2.93
After 30-Aug.	786	118	15	2.94
Calculated Value	800	133	16.7	2.7

4.4 Revamp of Liquefied Gas Desulfurizing tower

Requirements for packings in liquid-liquid extraction are different from that in vapor-liquid mass transfer. Mass transfer in liquid-liquid extraction is undertaking between droplet groups in continuous phrase and dispersion phrase, requesting the packings capable of dispersing droplets well and preventing them from accumulating. A special random packing for liquid-liquid extraction developed by Qinghua University has been successfully applied in industrial extraction column.[11] However, random packings are very difficult to meet more and more needs for liquid-liquid extraction column revamp for their time consuming manufacture and troublesome installation and unloading. Combining individual advantages of structured and random packings, Petroleum Refining Department of Chemical Engineering Research Center of Tianjin University developed a new kind of special structured packing for extraction. As a structured packing with self dispersion function, it retains the high capacity advantage of corrugated sheet metal packing and inherits the re-dispersion feature form Mini Cascade Ring (CMR), making the disperse phrase in best dispersing status all the time. Technology of structured extraction packing matched with new liquid disperser was employed in the revamp of liquefied gas desulfurizing unit of Fujian Petroleum Refinery and worked well.

The original designed capacity of liquid-liquid extraction desulfurizing unit in Fujian

Petroleum Refinery was 159,600 t/a (19.95 t/hr). The sieve trayed desulfurizing tower was of 1600 mm I.D. and 0.6% tray opening. The lowly designed capacity of liquefied gas desulfurizing unit did not match the demand by the 1.4 Mt/a RFCCU and the catalytic oxidation sweetening unit (24.903 t/hr). Its narrow operating flexibility can not keep normal production, for instance, when LPG throughput fluctuated within 19 ~ 25 t/hr, sever entrainment of both liquid phrase and of amine in LPG took place to disqualify the products. The only solution to maintain normal production was to reduce capacity or pump out a part of sulphurous liquid hydrocarbon to sulphurous LPG tanks. Hence, the revamp to desulfurizing tower was unavoidable to assure that liquefied gas desulfurizing unit matches RFCCU and gasoline catalytic oxidation sweetening unit.

The revamp included:

- Removed all 15 valve trays in the tower and then packed with extraction structured packing developed by Petroleum Refining Department of Chemical Engineering Research Center of Tianjin University. The packing combined the common features of corrugated sheet metal packing and CMR, that is, high capacity and efficiency, low pressure-drop and self-dispersion.
- Bottom LPG disperser was altered to a special tube-trough type. The initial distribution is of most importance for LPG as disperse phrase in the tower. The new tube-trough disperser is the development from the orifice tube distributor. Not only does this disperser give highly uniform distribution of liquid drops sprayed from it and overcome the contradiction between velocity and disperse quality, but also does it possess re-disperse function and allow the single inlet flow spreading into multiple streams. Besides, it has the advantages of large turndown ratio and low entrainment.
- The top inlet single tube distributor was changed to orifice tube distributor to ensure steady inlet flow and uniform distribution of the solvent, and also lowered the hole velocity for reducing entrainment and back mixing in the tower.
- A high performance bed of corrugated sheets was installed both under the bottom and above the top distributor to assure the separation of droplets of LPG and DEA (the solvent) for prevent entrainment in both sides.

The LPG desulfurizing tower started its formal operation in February 6, 1994 after completion of the revamp. Operation in past two years revealed that the tower has been running steadily and showed very effective. Testing data to the tower on May 8, 1995 was compared with operating ones, with the results listed in Table 7. Operating data in Table 7 indicate that:

- LPG throughput was largely raised, with average 25 t/hr and maximum up to 27 t/hr, increasing 31% or maximally 42% than before the revamp, as well as the flexibility and adaptability of the tower enhanced.
- Amine entrainment phenomenon was completely eliminated. At low throughput case before revamp, amine was so seriously entrained in LPG that normal operation was difficult to keep when throughput up to 20 t/hr. The phenomenon seldom happened after revamp even at high throughput.

● LPG desulfurizing effect was improved. Slightly unsteady operation or rising in H2S concentration in LPG inlet often resulted in disqualification of H2S concentration in LPG outlet before revamp, with only 85% being qualified. Elimination of entrainment after revamp bettered LPG desulfurizing effect jointly. In case of inlet LPG H_2S concentration ranging between 2,000 and 34,000 mg/Nm^3 and H2S concentration in lean oil as high as 3.79 g/l, the outlet LPG H_2S concentration qualified steadily over 98%.

Table 7 Operation Data Contraction of Before and After Revamp Showing Revamp Effect

	Before (1993)				*After (1994)*					*Test on*
	May	*Oct.*	*Nov.*	*Dec.*	*Mar.*	*Aug.*	*Sep.*	*Oct.*	*Nov.*	*5/8/95*
Inlet H_2S Conc. mg/Nm^3	3000	150 0	16000 ~ 25000	5000~ 13000	2000	11000	10000	7500~ 34500	3000~ 15000	3124
Outlet H_2S Conc. mg/Nm^3	0.005	0.03	>20	>20	0.003	0.054	0.018	0	0.017	0.0355
Capacity Range t/hr	18	20	21	22	23	22.5	15.8	18	15	19.54
Capacity t/hr	17~2 1	17~ 24	17~24	21~27	21~26	19~26	12~16	16~20	14~22	18.24
Amine in LPG t/hr	12	11	8	12	0.59	0	0	0	0	0
Pressure MPa	1.2	1.20 7	1.2	1.205	1.25	1.25	1.25	1.25	1.27	1.09
Recycle Amine t/hr	9	14	12	12	10	14	12	17	15	16.3
Lean Oil Conc. wt%	10	7.32	9	9.5	6.87	8.9	9.8	7.4	8.7	11.1
H_2S Conc.in Lean Oil, mg/Nm^3	1.88	1.9	2	2.17	3	3	3.2	2.8	1.9	3.79
Lean Oil Temp	40	38	39	38	40	38	36	37	35	41.3

References

[1] UPDATE., Chem. Eng. Process, 98(8), 24, 1994

[2] Koch Engineering Company Inc., FLAXIPAC Structured Packing Systems, Bulletin KFP—4.

[3] Norton Chemical Process Products Corporation, Intalox High-Performance Structured Packing, 410039/0694, ISP—2, 1994.

[4] JAEGER Products Inc., Metal MAX-PAK™, Product Bulletin 500, U.S.P. 4674934, 4710326.

[5] Liu Naihong, Application Handbook of New Industrial Tower Packings. Tainjin, Tanjin University Press, 1994. (In Chinese)

[6] Zhou Wei, Liang Taian and Du Jianting. PRC Patent ZL 95217757.9.
[7] Zhou Wei, Liang Taian and Du Jianting et al. Petro-chemical Equipment. 1998, 27(3):8~10.
[8] Zhao Ruwen , Li Xingang et al. PRC Patent, 89208579.7., 1990.
[9] Li Xingang, etc., Petrochemical Design 1994, 11(4):24~32. (In Chinese)
[10] Sun Jinsheng, Li Xingang et al.. Petroleum Refining Design. 1996, 26(1):19~21. (In Chinese)
[11] Fei Weiyang, Wen Xiaoming et al. Chem. Eng. 1995, 23(2):24~27. (In Chinese)

85a Mellagrid 64.X testing at FRI

Mark Pilling

The paper will present the results of the Mellagrid 64.X testing that was conducted at Fractionation Research, Inc. (FRI) in 2004 as a Category 1 test. The packing was tested at three standard FRI test systems/pressures. Results showed that the packing had exceptionally high capacity, exceeding the FRI calculated system limit. The FRI data will be compared with other grid and high capacity device data. The impact of the test data on commercial applications will also be discussed.

85b Identify Liquid Maldistributions in Packed Distillation Towers by CAT-Scan Technology

Simon Xu, William Mixon

Maldistribution of liquid in packed towers can impact dramatically mass- and heat-transfer efficiencies of distillation, absorption and stripping operations. Latest development in gamma computer-aided tomography (CAT-Scan) has made it possible to obtain an image of cross-sectional density distribution inside packed beds. Further work is needed for characterizing the scale of maldistribution from the CAT-Scan images. In this paper an image analysis approach is developed for assessing the maldistribution from the CAT-Scan images. Cases studies for diagnosing liquid maldistributions in industrial packed towers will be presented.

85c A New Method to Assess Liquid Distributor Quality

Lothar Spiegel

Liquid distributors are a key component to optimal performance in packed distillation columns. The most important requirement specifications are still the drip point density and the maximum standard deviation of the flows from the outlets. For modern baffle plate distributors, however these are no appropriate requirement specifications, because the liquid is delivered to the packing in form of drip lines rather than drip points.

This problem can be eliminated by analyzing the liquid distribution below the first layer of packing. The ratio of wetted to total geometrical surface is used as a quantitative measure (wetting index) to describe distributor performance. It allows to compare different distributor systems e.g. those with conventional drip points and those with drip lines. The wetting index depends on the geometry of the outlet system, the orientation of the distributor in relation to the packing and the structure of the packing underneath. This will be demonstrated by examples.

85d Hyperfil® Knitted Mesh Tower Packing – A Pioneer in the Past/A High Efficiency Leader for the Future

Mark Neuman, Martyn Cooling

Since pioneering the concept of fabricating a high efficiency packing as vertically laminated ribbons/sheets fastened together into segments that can pass through a manway and that rotate their orientation in a tower from layer to layer, Hyperfil has been in the forefront of mass transfer technology. Data is given showing how this element-to-element orthogonal re-direction of the capillary streams formed in the packings high, 1200 m2/m3 to 16,000 m2/m3 surface area leads to linear scale-up from 6" (150 mm) diameter and up. Forty years of pilot and field capacity and pressure drop data are shown on GPDC interpolation charts for several different styles of Hyperfil, while their efficiency data is presented as a function of C-Factor. High performance is also shown in the polymeric materials of polypropylene, PVDF, Teflon®, and dual components with any metal that can be draw into a fine wire, thus making knitted mesh packings the choice for availability and economy in corrosive mass transfer applications. Case histories are given demonstrating how this family of high efficiency tower packings has greatly benefited customers' bottom line.

85e Advances in Styrene Fractionation with INTALOX® Packed Tower Systems: Part 2 - FLEXIPAC® HC® Structured Packing

Gail Hausch, Izak Nieuwoudt, Randall A Sommerfeldt

Structured packing has been used commercially in styrene service for over two decades. New design advances continue to be developed to meet ever-increasing process demands and provide superior tower performance. INTALOX® Packed Tower Systems by Koch-Glitsch combine high performance FLEXIPAC® HC® Structured Packing with state-of-the-art high performance internals tailored to meet specific process objectives. Driven by recent styrene projects, technology improvements have been developed and implemented.

These design improvements were implemented in a recent world- scale EB/SM column revamp with FLEXIPAC® HC® Structured Packing. Packing operating efficiency and pressure drop data are presented for this large scale commercial column and compared to total reflux distillation test data performed in smaller diameter columns. The same type and size structured packing has been tested both in the Koch-Glitsch laboratory column and at Fractionation Research, Inc. (FRI) under similar vacuum conditions. Comparison data from these three sources demonstrates that packing performance can be replicated in commercial installations of large diameter columns that have large numbers of thoeretical stages.

85f Modeling and Validation of Random Packing Hydraulics

Florian Kehrer, Lothar Spiegel, Peter Choo

With an increasing dependence on vendor supplied design tools for the calculation of packing hydraulics, the basis of this software needs to be reliable. Currently however, different vendor supplied design tools offer different results for some very similar packing types.

This paper sets out to determine a reliable model for the hydraulic calculation of I-Rings® (IMTP® equivalent). The full range of I-Ring® sizes from #15 to #70 was tested with the air-water system at medium to high liquid load ranges in the Sulzer Chemtech facilities. In addition, distillation tests with standard test mixtures under total reflux condition were conducted. The results of the hydraulic modeling and validation exercise are presented.

Finally, these data will be compared with different vendor supplied design tools.

85g Comparison of Packings for Absorption with High Liquid Loading

W Y Fei, X Y Song

The hydrodynamic and mass transfer of Plum Flower Mini Ring (PFMR), Pall Ring, Intalox Saddle and Mellapac were studied and compared in a 600mm diameter column with air-oxygen-water system over a wide range of liquid loads (from 10 – 220 m3/m2 hr). It was shown from the experiments that the PFMR had much lower pressure drop, larger throughput and higher mass transfer efficiency than Pall Rings and Intalox Saddle. Furthermore, the mass transfer efficiency and volume efficiency of PFMR were higher than Mellapac with similar specific area under high liquid loading.

In order to improve our understanding on the experimental results, CFD stimulation and PIV measurement were carried out. It is shown from the results that the ratio of height/diameter of ring packing has significant influence on packing performance.

Case studies are provided to demonstrate the advantage of PFMR.

85h Comparison of liquid distribution performances of conventional and high capacity structured packings

Zarko Olujic

This paper presents/discusses the results of a comparative experimental study on liquid distribution behavior of conventional and high capacity corrugated sheet structured packings, in conjunction with different distributor designs and variations in influential variables, including the density and layout of irrigation points, the liquid and gas load, the bed height and the type of imposed initial maldistribution pattern. The liquid distribution profiles including the information on the extent of wall flow were obtained using the TU Delft column hydraulics simulator with an internal diameter of 1.4 m. The packing tested were Montz-pak B1-250 and B1-250M, the test system was air/water, at ambient conditions. An empirical characterization factor is introduced which is less dependent on grid size than the so-called maldistribution index. Strikingly, the high capacity packing exhibited a superb liquid distribution performance and, in general, it appeared that it is more important to distribute irrigation points uniformly than to have uniform flows from them.

86b Solution Foaming in Gas/Liquid Treating Towers

Steve Von Phul

Foaming has gained the distinction of being the number one cause of gas/liquid contact and stripping tower upsets. However, foams do not occur spontaneously. They form when gases and liquids are mixed in high shear environments. Trayed towers are designed specifically to produce fast breaking foams commonly referred to as "froth". Froth provides the mass and heat transfer area responsible for the proper function of these columns. If these towers generate foams in order to work properly, why is foaming considered such a ubiquitous problem?

Chemical antifoams are used by most plants operating gas/liquid treating towers to mitigate foaming problems with varying frequency. Yet, information regarding the chemical nature, function, and use of these additives is difficult to find in the petroleum industry literature.

This article describes, in practical terms, the fundamental causes of solution foaming, and factors responsible for foam stability. Positive and negative affects of using antifoams to mitigate foaming problems will also be discussed. Field results from actual foaming studies will be presented.

Foaming has gained the distinction of being the number one cause of gas/liquid contact and stripping tower upsets. However, foams do not occur spontaneously. They form when gases and liquids are mixed in high shear environments. Trayed towers are designed specifically to prduce fast breaking foams commonly referred to as "froth". Froth provides the mass and heat transfer area responsible for the proper function of these

86c Retrofit of a H_2S Selective Amine Absorber Using Mellapak Plus Structured Packing

Geronimo Sixtos, Glenn Shiveler

A major natural gas amine sweetening plant retrofitted its 48-inch diameter Amine Absorber from valve trays to high performance Mellapak Plus™ structured packing to increase gas capacity, and to selectively remove H2S and improve slippage of CO2 using generic-MDEA amine solvent. The retrofit of the Amine Absorber was a success, and it achieved the primary objectives of improved gas handling capacity and increased H2S selectivity. During six months of operation, the unit regularly achieved 80% slippage of CO2 while selectively removing H2S to pipeline specification. This report provides an overview of gas sweetening plant operation following the retrofit with performance data under variable inlet feed conditions. The report also provides observations regarding changes in amine quality, issues with amine degradation, amine salt contamination, experience with minor amounts of fouling and the utilization of an anti-foam agent.

86d Radioisotope Technology - Benefits & Limitations in Troubleshooting Packed Beds in Vacuum Distillation

Paul Hewitt, Scott Vidrine

The use of radioisotope technology in Chemical, Petrochemical and Refinery Process Plant mass transfer vessels has continuously been developed over the past 50 years to assist engineers in troubleshooting process problems, optimize production and plan shutdown schedules to minimize plant down time. Many advances in data acquisition and processing ability now allow critical information to be provided quickly to ensure the most efficient use of plant equipment.

With the advancement of data gathering and handling it is important to note that fundamental physics behind each and every measurement involving radioisotopes has never changed. As with any technology there are many advantages in its use. However there are also limitations that a process engineer must fully understand so that a possible problem is not overlooked or simply discounted based upon results generated.

Mass transfer design can typically be split into two main categories; trayed and packed beds. Packed beds are becoming increasingly popular. In the case of a trayed tower, the open area is typically less than 10% whereas the open area in a packed bed is greater than 50%. Additionally, packing tends to have a lower liquid holdup than trays. These characteristics combine to result in pressure drop for a theoretical stage being lower in packing than in trayed systems. This reduction in pressure drop can have a significant impact on tower and plant economics.

As with all situations, packed towers bring their own particular challenges. As an example, the same open area and reduced liquid holdup that benefits pressure drop may result in lower liquid and vapor velocities, and therefore, an increased susceptibility of packed beds to accumulate solids in systems prone to polymerization or precipitation. Another common issue surrounds mechanical integrity. Packed bed towers are more susceptible to damage as a result of pressure surges as compared to trayed vessels. Finally, a critical consideration in the effective operation of a packed tower is the mechanism for fluid distribution. Poor vapor or liquid distribution can result in a significant efficiency reduction. In fact, the necessity of uniform distribution increases with the number of theoretical stages in a bed. This results in tower internal designers limiting bed length and redistributing process materials at specific and regular intervals.

With the increased use of packed beds the application of radioisotope technology to troubleshoot problems has become popular. This paper, with the aid of case studies, describes the general principles of the radioisotope diagnostic technologies used in vacuum towers that allow operating characteristics to be measured. It shows the benefits that can be gained by using the technologies as well as limitations and in some cases the necessary strategy of radioisotope technology integration to successfully determine the source of a particular mass transfer problem.

86e Component Trapping in Distillation Towers: Causes, Symptoms and Cures

Henry Kister

This article explains the principles behind minor-component trapping, summarizes industry's experience, and provides guidelines for diagnosing and addressing the problem.

86f Scan/Simulation Integrated Diagnostics for Distillation Improvement

Simon Xu

Scan/Simulation Integrated Diagnostics™ (SSID) is an engineered approach towards problem solving that combines distillation column scanning with the latest developments in simulation software – for a truly comprehensive column diagnosis. Simulations assume mechanical integrity at the outset. They are not able to make allowance for damaged trays or packings.

A column scan can quickly identify any of a number of problems, ranging from damaged or missing trays to fouled packings or packing distributors. On the other hand, a scan can only detect problems that can be observed as a density difference. Therefore, a control system malfunction might go undetected with a conventional scan, but can be easily detected with a process simulation (by checking material and heat balances).

SSID can be applied to all liquid/vapor contacting systems (such as distillation, absorption, and stripping processes) and most tray and packing designs. A column scan working in conjunction with the most advanced simulation technology commercially available is the most powerful strategy for troubleshooting, revamping, or optimization. Case studies are presented.

86g Trapped Components in Distillation Columns: Sometimes Changing Operating Conditions Can Solve Your Problem

Soundar Ramchandran

Minor components trapped in distillation towers can sometimes be a source of major problems. In some cases, the minor components may be reactive inside the column environment and react to form higher molecular weight species and polymers that may not be soluble in the liquid phase in certain sections of the column. The presence of solids inside the column often results in unpredictable vapor-liquid contact behavior and results in operational headaches that could include adverse effects on product quality and increased maintenance cost and downtime for plants. Often solutions explored for such a problem comprises of methods that enable operations to deal with solids after they have formed (i.e., modification of trays and internals that deal with solids more easily) and/or minimizing the formation of the polymer species (i.e., injection of additives to inhibit the reactive species or disperse the polymers). Typically, these solutions tend to be expensive and prove to be somewhat of a "Band Aid" for the real problem. One option is to take a close look at the column operation and explore ways in which the minor component can be purged from the process. A case study in which the minor component trapped in a column was purged effectively by a very slight change in operating conditions will be discussed in detail.

86h Troubleshooting an Ethylene Quench Water Column

Donna Fruge, Randall A Sommerfeldt, Timothy M Zygula

The authors will present the case of a poorly performing ethylene quench water tower. The column was experiencing severe pressure drop problems and losses in heat exchange. The packing used in the initial design of the column was random packing with spray nozzle distributors. The packed bed in this column was damaged and replaced with grid type packing. The authors will detail the steps taken to troubleshoot and revamp the column. An overview of the revamp design for the column will be detailed in the article. The authors will also present operational data from the quench column. The data that will be presented includes number of heat transfer units, temperature and pressure profiles. The heat transfer capacity of the column before and after the revamp will be presented.

86i Strategies of Successful Distillation Equipment Revamps Part 1: Determining Existing Distillation Equipment Capacities

Daniel R Summers, Karl Kolmentz, Peter W Faessler, Ryan Pitt, Wai Kiong Ng

Successful unit operation revamps are those which have a high return on investment and low life cycle cost. Life cycle cost may include original capital, energy, operation, and maintenance cost over a typical life span of 10 years. For simplicity, some life cycle cost evaluations only include originally capital and energy cost. Distillation equipment is typically greater than 25% of total original capital cost and normally the easiest to upgrade.

Operational Plants have an original capital value per yearly production. For example, a new ethylene plant ISBL (inside the battery limit not including utilities and tank farms) might have a cost of US$ 400.00 per KTA (thousand tones per annum). A new 300 KTA ethylene plant might cost US$ 120,000,000.00. To be a cost effective revamp, the revamp should have less cost than the cost per ton of a new plant. With the proper selection of key items this can easily be accomplished, while also improving the life cycle cost.

Successful revamps generally have several key steps. The first step is a high load test to accurately determine the existing distillation equipment's available capacity. The second step is to reconcile the data gathered in the high load test by a heat and material balance. This normally utilizes distillation software such as Pro II. The third step is with the reconciled data, rate the existing equipment verses it's designed and target values. The fourth step is to select the key items to be upgraded with the maximum return on investment.

Planning and executing a high load test is a critical part of any successful revamp. It can identify and / or eliminate key equipment to be replaced or retained. Completion of an accurate high load test can simplify the steps which follow, with the ultimate result – a successful revamp.

American Institute of Chemical Engineers
3 Park Avenue, New York, NY
10016-5991

ISBN: 978-1-60423-465-7